Integrated Logistics Support
The Design Engineering Link

IFS

Integrated Logistics Support

The Design Engineering Link

Walter Finkelstein
and
J.A. Richard Guertin

IFS Publications, UK
Springer-Verlag
Berlin · Heidelberg · New York
London · Paris · Tokyo

British Library Cataloguing in Publication Data
Finkelstein, W. and Guertin, J.A.R.
Integrated logistics support
1. Materials management. Logistics
I. Title
658.7

ISBN 1-85423-011-5 IFS Publications
ISBN 3-540-50649-7 Springer-Verlag Berlin Heidelberg New York Tokyo
ISBN 0-387-50649-7 Springer-Verlag New York Heidelberg Berlin Tokyo

Phototypeset by Area Graphics Limited, Letchworth, UK.
Printed and bound by Short Run Press Ltd, Exeter, UK.

CONTENTS

PREFACE

This book has been prepared to assist the various functions within a corporation, specifically those doing work in the integrated logistics support (ILS) field, with the material necessary to ensure that ILS is properly integrated among all of its support functions. It identifies the interfaces required to ensure that, as equipment is designed, each of the appropriate functions under the umbrella of ILS is properly addressed.

We hope that the presentation of this material will give the necessary insight to ensure the proper functional flow of work that is necessary to support a total acquisition life cycle. The book itself has been designed to be applied to both military and commercial approaches. Our intent is to ensure that integrated logistics support is applied commonly to both avenues of industrial endeavour.

Some of the terms and definitions that are applied to military operations may not be necessarily applicable to operations in the industrial complex. We have tried to ensure that parallelism occurs between both areas, military and commercial, so as to bridge the specific functional elements discussed. We hope it will lay out the groundwork to show how each of the ILS elements becomes a major player in the total life cycle of a product. We have prepared a roadmap that, when properly used, will establish the baseline for all activities within the ILS function as it relates directly to design and programme management.

The book begins with the elements that we believe to be most essential to the accomplishment of ILS. They are complemented by flow diagrams illustrating the approach that we have just discussed. We believe that the combination of narrative and illustrations will simplify the information presented so that people just beginning to become familiar with ILS will find it very simple to follow.

August 1988

Walter Finkelstein
Richard Guertin

ACKNOWLEDGEMENTS

I would like to extend my sincere thanks to my wife Kay and family for all of the assistance they have given me in the preparation of this book. Also, I would like to thank my co-author, Dr Richard Guertin, without whose contributions in the area of the commercial aspects of integrated logistics support this book could not have been completed.

Walter Finkelstein

I also extend my appreciation to my wife Mary and family for the support, patience, understanding, and many suggestions and assistance provided me in support of this task. Further, my thanks are extended to Walter Finkelstein, who throughout this effort, maintained the best of a collegial relationship which evolved into a friendship.

Richard Guertin

Chapter 1 INTRODUCTION

The term 'logistics' is probably one of the most convoluted terms used in business, industry and government today. Everyone seems to think he knows what logistics is, but when it comes to defining it, there are a myriad of explanations. Our intent is to present the study of logistics for the journeyman in capsule form, with the goal of attempting to make the picture of the logistics field more unified, understandable, useful and profitable.

Before getting too involved in the content of the book, some history on the origin of logistics is necessary, with an explanation of how it was implemented in the US Department of Defense (DOD), and why it should be implemented in the private sector.

The word 'logistics' comes from the Greek *logistikos,* meaning 'skilled in calculating'. The word *logista* first appeared as a title applied to a rank of military official during the Roman and Byzantine empires, but was not really used in the true military sense until the eighteenth century. Antoine Henri Jomini, Napoleon's biographer and the leading military theorist in Europe during most of the nineteenth century, finally fixed logistics in military parlance as a term referring to military administration, though not necessarily embracing the entire field. In his celebrated *Précis de l'art de la guerre,* he built a theory of the art of war on a trinity; strategy, 'grand' tactics and logistics. His main concern was not the actual art of logistics, but of military management as a whole. Despite the enormous influence of Jomini's writings on military thought during the middle of the nineteenth century, his few remarks on logistics drew little attention. Its strict interpretation was applied to the supply, quartering and movement of troops. Logistics was not truly introduced until late in the nineteenth century, when Admiral A. T. Mahan introduced it into the US Navy. His interpretation expanded the term to include the process of industrial mobilisation.

All through World War I, the word logistics was little used and generally applied in its narrowest sense. Not until World War II did logistics come into vogue. The term was used indiscriminately, and it tended to lose its more restricted sense. At the conclusion of the war, the Army broadened the term, and it was finally published in terms of its own function, to include:

- Procurement.
- Storage and distribution of equipment and supplies.
- Transportation of troops and cargo.
- Construction and maintenance of facilities.
- Communications.
- Care of the sick and wounded.
- The induction, classification, assignment, welfare and separation of personnel.

Though logistics now appeared to encompass a wider range of activities, it still had strong conservative and traditional meaning.[1] The problem that began to emerge was a lack of concise definition of precisely what logistics should encompass and what its applications should be. It was not until McNamara's term as Secretary of Defense in the early 1960s that the DOD recognised that a means to coordinate the implementation of all of the different operations would be appropriate. This action was part of the so-called System* Analysis thinking that spawned the Planning, Programming and Budgeting System and was caused by the explosion of technology. Rapid changes began to take place, first with the introduction of the transistor and then with the advanced applications of the computer. Programs started to increase in complexity, and the funding required to support research and development programmes was becoming scarce.

The DOD realised that if this situation was not placed under some sort of control, they eventually would be faced with the prospect of a burgeoning logistics budget. The integrated logistics support (ILS) procedures perceived by the DOD focused more attention on the independent, structured elements identified above. The list was expanded to include other elements that were directly related and could be used as a means to develop a cohesive, well-orchestrated management programme. These additional factors resulted in a new perception that not only integrated the logistics elements, but also assisted in the integration of logistics with the system engineering function, which was sorely needed. The principal document through which all of this occurred was first published in 1964. The publication took the

[1] *Encyclopaedia Britannica* (1959) vol. 14, pp. 334–334A.

* The author's definition of a system is the combination and integration of all parts to make up the whole.

form of DOD Directive DoD-D-4100.35, entitled *Integrated Logistics Support Guide*.[2]

The integration of these elements under a single umbrella took some time to gain momentum. As a result of years of patience and continual effort by many people, both inside and outside of the DOD, the following events occurred.

In May 1969, Admiral Galantin, United States Navy (USN), then the Chief of Naval Material, stated that:

> The small cost of ILS planning, less than one percent of the first article acquisition cost, is well worth the investment in the light of the large support cost: i.e., six to ten times the acquisition cost. Based on past experience in other programs and results of initial ILS efforts in current programs, ILS planning is expected to lead to reduction in life cycle support of $100 to $200 million per major ship program.[3]

This statement of policy ensured that the US Navy would become a major player in the ILS arena. It was only one of many statements issued at that time that provided the thrust to ensure that logistics, as a distinct field of specialisation, got off to a positive start. The strongest and most profound statement of support for ILS was made to the Congress of the United States. It provided the necessary impetus to reinforce the need to continue the drive to full implementation of ILS planning.

The following is an excerpt from a statement made by David Packard, then the Deputy Secretary of Defense, before the House Subcommittee on DOD Appropriations on selected aspects of the FY72-76 Defense program (18 March 1971).

> Integrated Logistics Support. It is clear that logistics support is a major portion of the total cost of a weapon system. We have established an Integrated Logistics Support Program to assure better management of weapon system logistics. This ILS will include . . . Maintenance . . . Supplies . . . Support and Test Equipment . . . Technical Data . . . Facilities . . . Maintenance Personnel and Training . . . and Maintenance Management information. [Logistics Funding and ILS Management were added later.] ILS is now a part of the overall acquisition process. The ASD (I&L) and DDR&E have developed a directive covering ILS and this function is now included in the Development Concept Papers (DCP) and is routinely considered as a portion of the DSARC meetings.[4]

The document referred to in the Secretary's comments was published as a

---

[2] US Department of Defense (1964) *Integrated Logistics Support Guide*, DOD Directive DOD-D-4100.35.

[3] Admiral Galantin, US Navy Department, 1969.

[4] US Congress, House Subcommittee on Department of Defense Appropriations, 94th Congress, 1st Sessions, 18 March 1971.

revision to the DOD's *Integrated Logistics Support Guide*. The modifications that were included did not just improve the application of ILS, but also structured the elements defined by Mr Packard into a cohesive functional element that started to take on its own characteristics. The other major change was that the directive was established as a guide for all military agencies.

The fundamental concept of ILS was further refined to mean a composite of the elements required to ensure the effective and economic support of a system/equipment at all levels of maintenance for its programmed life cycle. The intent of the Secretary's statement was that all areas of expertise should cooperatively participate using an integrated approach to successfully deploy a fully supported system. In this manner, logistics strategies would be combined with system engineering techniques, including cost effectiveness studies, to obtain a supportable and affordable system. These studies and their results are used to determine the relationship between hardware design characteristics and logistics support requirements. Following this philosophy, concentration is on the system engineering level so that all of the mission requirements can be measured and the proper design techniques can be employed at the earliest possible time in the product's life cycle.

From a macro point of view in terms of the economy and external environment, logistics contributes to the economic value of products and services by adding to their form, time, place and possession utility. In addition, the economic impact of logistics affects economic development and specialisation, types of goods and services, price effects and land values.

In the United States in 1986, logistics costs exceeded $400 billion, or 11.1 per cent of the Gross National Product.[5] These figures represent only inventory, transportation and physical distribution costs.

Logistics is the key to a firm's competitive advantage in terms of cost leadership (lowest cost production of goods and services, differentiation and uniqueness of the business enterprise according to feelings of customers) and focus (a situation when a firm strategically sets its sights on having a competitive advantage).[6]

Components integral to the logistics field are defined differently, depending upon the field of endeavour in which an individual is working. From the commercial aspect, logistics can be viewed as production operations and physical distribution and transportation. However, efforts are being made to integrate the terms being applied in the military field into the commercial environment. Many of the terms and definitions applied to the military aspect of logistics are now being adopted and integrated into the logistics field as seen through the eyes of the private sector. Regardless

---

[5] Coyle, John J., Bardi, Edward J. and Langley, C. John Jr (1988) *The Management of Business Logistics* (St. Paul, MN: West Publishing Co.), p. 4.

[6] Ibid., p. 4.

of the approach being taken, logistics interfaces with the organisation's infrastructure in marketing, technology development, planning, finance, accounting, quality management, human resources, development and management, personnel and services to maintain the value and usefulness of the product.

One of the major elements that must be maintained and controlled is the logistics interface with the systems engineering function. This marriage must be continued and a total team effort must be established so that the interactions among all of the functions early in a program are properly controlled. The marriage of systems engineering and logistics ensures that no single action will be allowed to stand alone, regardless of which functional area performs the action. This is true for engineering and all of the supporting functions necessary to field or distribute a specific product or military system. Proposing changes to a design without consulting the other functions and without determining the impact the change will have on the supporting functions will have a rippling effect on all aspects of the follow-on support elements. The problem finally comes to rest with the technician who is ultimately responsible for repairing the product. Therefore, performance capability, modularity, commonality objectives and system availability must be maintained against the specific needs of either the military or the consumer, measuring all of this against cost.

To ensure that logistics maintains its appropriate profile within an organisation, an ideal logistician is a generalist who makes significant contributions to the firm. Logisticians can be considered troubleshooters, auditors, expeditors and coordinators. Increasingly, many firms are beginning to establish logistics management divisions that have overall broad logistics responsibilities.

Logistics, or ILS, interfaces with all of a firm's functional elements and is essential because of the bridge logisticians provide between these functions. The result is reduced conflict of objectives, limitation of redundancy, optimum trade-offs that take into account the sequence of decisions, combination of competing functions with removed cross-function coordination, reduced cost, greater productivity and an increase in corporate profit.

ILS spans the spectrum of all of the system support activities, from concept formulation and development through the entire life cycle process to obsolescence planning and waste management. The goal of ILS is to provide a system's maximum system effectiveness with all of the required logistics support throughout a system's life cycle. ILS must be applied throughout the process to ensure that the system meets the operational requirements of the user. Under this concept, the importance of trade-offs and operational and support requirements should be recognised from the beginning. These trade-offs must be balanced against all of the other functional efforts being performed to satisfy the ultimate goal – fielding or deploying a piece of equipment or an end item that satisfies the user's needs. The trade-offs must be performed to ensure proper coordination and control, with the ultimate result being a compromise among all functions that satisfies the minimum essential requirements – fielding an operationally supportable piece of equipment at the minimum life cycle cost.

In addition to integrating support planning into the entire design and development process, ILS must also be integrated into the total support system; it is the composite of elements necessary to ensure the effective and economic support for its programmed life cycle. All levels of expertise should cooperatively participate, using an integrated approach to successfully deploy a fully supported system. The logistician works closely with the team to incorporate logistics strategies with system engineering techniques (including cost effectiveness studies) to obtain these ends. Thus studies and their results are used to determine the relationship between hardware design characteristics and logistics support requirements. When the baseline of any logistics element is changed or proposed to be changed, the effect on the total system must be formally considered and necessary adjustments must be made.

When applying logistics to acquisition, it is important to understand that logistics support is not an end in itself; it exists only to support the operation and use of the product. The practice and application of logistics support serve to facilitate the development and operation of the product. The support problem will vary according to their complexity and value. Planning and support must be individually tailored to satisfy the phase in the acquisition time cycle.

The first part of the logistics problem in the acquisition process, then, is the establishment of basic operational requirements. What do we want the product to do, and how should it operate? Programme, project and product managers must keep the operational purpose clearly in view during the earliest stages. These managers should recycle and refine their planning to determine the minimum threshold to be accomplished before full-scale development.

Once the basic logistics system characteristics and requirements have been formulated, managers must adequately state them to the design engineer throughout the systems engineering development cycle so that they can be used in the analytical and trade-off studies. Integrated logistics support is not intended to stand alone as a functional island. It must be integrated with all the other functions that together produce and sustain an operating end item or product or service. In the development of logistics support, these managers must ensure that logistics and design personnel cooperate. This cooperation should be widespread and integrated, as earlier indicated, throughout all the other appropriate portions of the corporation. Thus ILS must be accepted as an integral part of the total system engineering life cycle. The ILS concept in practice continues on through the production operation phases.

During the production phase (ideally), the entire effort should be turned over to the logistician, because in essence he is responsible for ensuring that all of the parts, support equipment, training, transportation, inventory control and other logistics elements, are established and properly integrated into the 'main'. This 'main' could be the corporation or a military establishment that will be responsible for maintaining and operating the equipment.

Therefore, the key to proper logistics support management is a systematic and orderly approach to early planning. Effective ILS requires the identification of precise systematic requirements early in the development phase of the product. Timely achievement of requirements is essential as the programme or product progresses during its life cycle, as is total responsiveness to change. This last aspect occurs during the full-scale engineering development and production phase and is carried on through the operational use, support phases and obsolescence of the product. (This will be further discussed in Chapter 6, covering the efforts of the configuration manager and his role in ensuring that a product is properly managed and controlled to satisfy the total responsiveness to change.)

To be able to manage all of the functions just described, the logistician should develop a document through which he functionally coordinates the efforts under his domain. This document is normally called an integrated logistics support plan (ILSP). This plan describes what is going to occur, who is going to be involved and when it is intended to be accomplished. It also establishes the necessary coordinating tasks that must be accomplished by each of the supporting functional managers under the logistician's control. This document becomes his bible, as the programme management plan is the project manager's bible, to ensure that the appropriate functions are produced and coordinated so that they interact and are properly supporting the equipment being designed. Most ILS managers consider the ILSP to be their mandate.

The first generation of this document is usually considered to be the requirement statement for all follow-on logistics efforts. This plan should specify the tasks to be performed during the early phase in the product's life cycle. Further refinements are then incorporated as the design matures. At the completion of the design, this document is converted into a support planning document, which eventually is distributed to all of the users. The logistician considers this plan to be a 'living' document, because it does not remain static: it takes on a life of its own, changing in accordance with the maturity of the system. It refers to all supporting plans and procedures approved by the logistics manager during the early life of the product. The support planning document lives in continuous application as long as the product is being used.

When all of the logistics elements are coordinated and properly sequenced, the true impact of logistics management becomes apparent. Results indicate increased quality, ultimate and optimal user satisfaction and higher profits to the firm. The key to success is that the logistician's participation occurs not only on an equal basis, but in a cooperative relationship with the other functions within the corporation or agency. Thus, the logistics support function has the ability to participate early in the decision-making process, and each decision can be measured in terms of the field of logistics. A corollary to this is the fact that engineers will benefit from logistics' participation to the extent that possible problems can be avoided. The logistics approach using a total team effort will encourage exchanges of ideas so that the design can be built in a minimum of time, without unnecessary costs, and at a low level of turbulence.

The early logistics effort must be confined to the development of inclusive but broad logistics plans and support. This effort should result in a road map of the specific steps to be taken and the time and detail of these steps. The planning and preparation of data packages (e.g. engineering drawings) must be deferred until the hardware configuration has been reasonably established. Detailed support planning, which is accomplished prior to the establishment of the basic configuration and is dependent on that configuration, is almost certain to require extensive rework to become valid and usable.

While the application of ILS is mandatory, every acquisition need not be planned to the same level of detail. What is intended, however, is that positive consideration of ILS must be ensured for major programmes. For lesser programmes, a review of the ILS elements should be performed. Tailoring these principles to suit the need of the acquisition at hand is of primary concern.

Chapter 2 LOGISTICS PLANNING AND MANAGEMENT

The basic objective

The basic objective of proper logistics planning and management is to achieve maximum availability of parts and personnel to ensure consumer acceptance of the product along with minimum life cycle cost of ownership. It is essential that these objectives be achieved, because without them logistics support will, in essence, be detrimental to the total design. Maximum availability of parts means that we have actually influenced the product. Life cycle ownership costs ultimately translate into the logistician's ability to create and implement an effective and timely support system meeting the requirements of the contract. Logisticians cannot perform these tasks by themselves. They must be accomplished as part of a team effort. Working with the design engineer and the other supporting functions, the logisticians will be capable of achieving the basic objectives.

Fundamental concept

Therefore, if logisticians have any intention of achieving these objectives, the fundamental purpose of ILS is to relate support to design and to use an engineering analytical approach rather than a rule of thumb to design logistics support subsystems into the hardware acquisition in order to ensure the timely achievement of objectives.

Corporate organisations

To integrate ILS at the highest levels of management within a corporation means it must be introduced at the senior vice-president or director levels. Direct control of ILS through this individual will give the necessary effective impact so that any problem that would occur between the functional areas of programme management and ILS can be readily resolved through proper lines of communication. Our object has been to define and illustrate a typical organisational structure that would be operating within any kind of organisation; one that is responsible to report to and be controlled by military-type contracts, or those that are necessary to support commercial ventures.

The organisational structure becomes the foundation from which all information flows. We not only look at what it means to schedule programming milestones, and to control the funds and quality aspects of our effort, but also consider what is required to ensure that the proper foundations are laid for achieving the ultimate goals. These goals are customer approval, a cost-effective support design configuration compatibility with other previously developed items, a programme devised and developed within support costs and a profit yield that is reasonable when integrated with ILS. Fig. 2.1 shows the relationships between effective ILS management efforts (schedule, money and quality management) to ensure that the ILS skills and the resulting efforts of ILS are fully achieved.

An effective ILS manager must be able to maintain three basic elements: schedule, money and quality. These are the rudimentary elements of ILS management. They are also the rudiments of good programme management. Therefore, the transfer from programme management to logistics management should be easy. The three elements of schedule, money and quality management will now be discussed from the ILS manager's point of view.

Schedule management

To ensure that the ILS elements are properly integrated, a schedule must be prepared in conjunction with the total programme, and realistic planning techniques must be applied. Visibility and programme monitoring are necessary. These two elements are accomplished through the use of scheduled management programme reviews. Problem resolution, on-time emphasis and fast reaction capabilities are essential if the logistics manager expects to achieve the goals set forth for the programme. These elements must be tied into the total programme so that as problems arise they can be easily translated into design solutions.

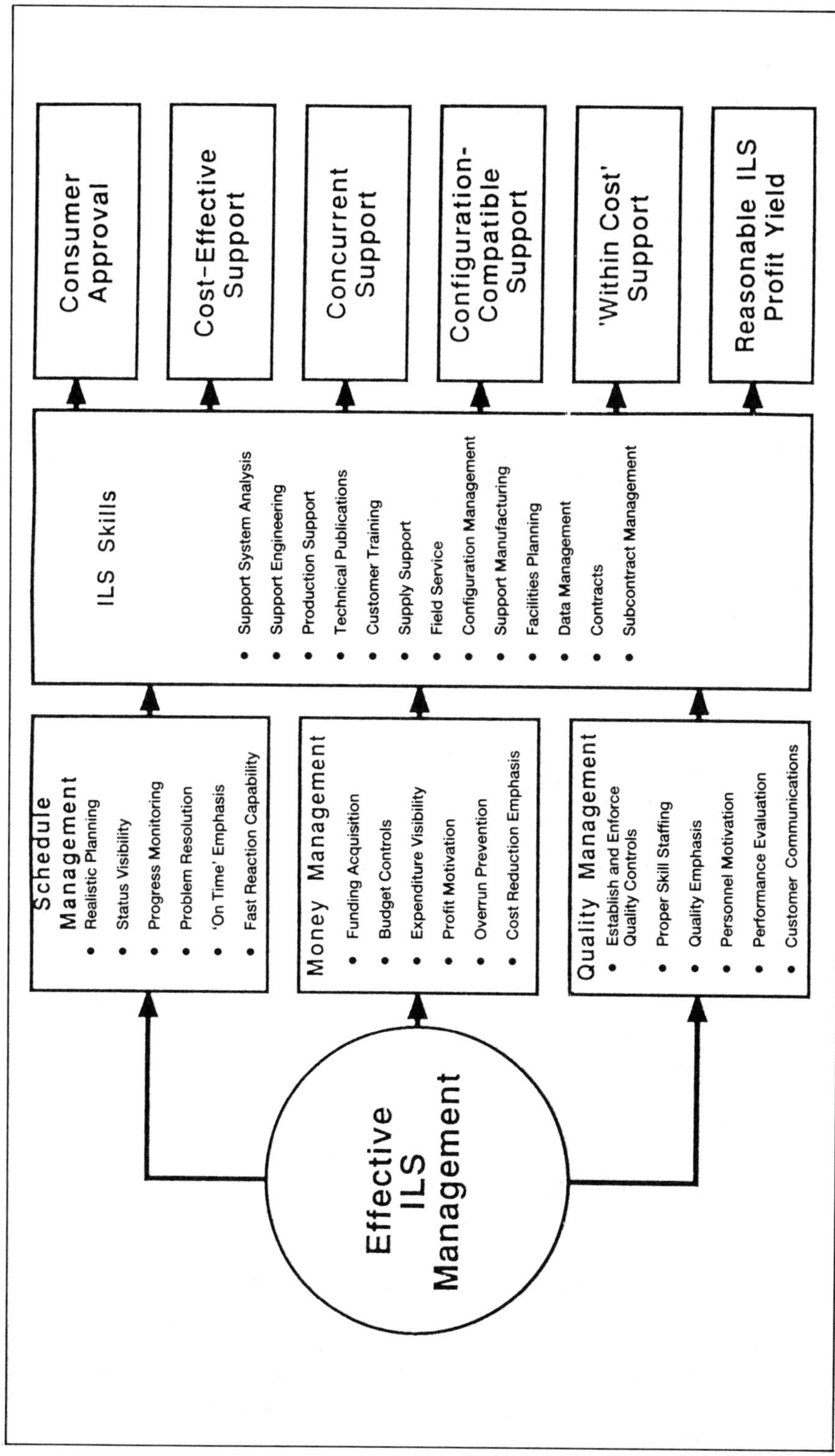

Fig. 2.1 Effective ILS management.

Fast reaction capability becomes another essential element in the total schedule management approach for the logistician. One should not delay, neither should one proceed beyond the design engineer, risking being criticised as aggressive, demanding in requirement and unsupportive of the total design. Therefore, one must make sure that interactions with the other functions are in consonance with the total programme effort.

Money management

Funding the acquisition is another problem. This is much the same as it is in any other programme, where the logistician often becomes the scapegoat, so to speak, when design engineering runs into financial problems. It must be assured at the onset of the programme that the programme manager will not cut into the logistics funding profile. The logistics manager must be in a position to ensure that, as the programme progresses, he has total control over the logistics budget. Expenditure visibility must be maintained. Logisticians must know what funds are expended and the degree to which it will affect the total design. Without this control, the programme manager may see fit to cut back and/or delete logistics support if he feels that ILS managers are not producing or carrying their weight as required by the programme.

Constant balance beween design and logistics has been discussed. This also applies to the amount of funds used to ensure that the logistics manager is able to maintain the necessary purview over what is going on as part of the total programme. That entails the application and control of his own financing.

Care must be taken not to over-run or even under-run the funding profiles that have been established for each one of the tasks described as part of the work breakdown structure. A funding over-run means that the logistician does not understand the programme requirements. An under-run implies that he has a full appreciation of the programme's requirements and has, in essence, padded it beyond the required safety factors. Both cases are dangerous and must be carefully monitored. History is a good teacher. The logistician looking at what has been accomplished in the past for the tasks to be performed on the next contract is essential if the logistician is to be able to bid his programme successfully. Using this technique, the logistician can accurately bid his requirements for all future contract needs. Being in a position to accurately identify the funds required to perform these tasks will ensure the logistics manager a relatively positive place in the programme management infrastructure.

The profit motivation then becomes the ability to win as many contracts as possible. The logistician has the ultimate role in attempting to identify or predict what his costs will be over a long period of time. Life cycle costing becomes an essential element in his domain. By applying a life cycle cost model, he can perform cost reduction analyses and trade-off studies, measuring the impact that specific logistics elements such as spare parts, test equipment, transportation and even training will have, if in fact they are properly identified and appropriate measures are taken early in the design. Any additional funds, even though they appear to be somewhat extravagant early in the design phase, could ultimately save the consumer much support money.

Quality management

Quality has always been and always will be the most important element of any function being performed for a design. Quality cannot be inspected into a product; it must be designed into it. The logistician and the effect he has on the design and the products that he ultimately delivers to the end user must be as high in quality as the design itself. If the technical manuals that are produced are low quality, the image of the corporation producing the product is tarnished. Therefore, in conjunction with the quality assurance group, the logistician must enforce the quality controls established by the quality assurance function.

Tasks essential to quality must be intuitively performed by the various members of this staff. The logistics staff must be appropriately motivated so that they will do the job right the first time. Repeat performances or inaccuracies in documentation lead to a total collapse of quality control in documentation. The greatest percentage of all of the work is rework. This is due to the lack of quality control involvement. Performance evaluation, therefore, must be measured in light of the quality aspects of a person's performance. Establishing the guidelines necessary to include quality assurance must be foremost in everyone's mind. Without this, performance to the requirements of the contract is not necessarily achievable. Quality plays an integral part in the total programme life cycle.

The quality management group within the logistics support environment is also the mechanism by which the logistician's values are communicated to the consumer, be it military or commercial. From the commercial point of view, the private industrial sector has a liability problem if a product does not perform to the level to which it has been described. If the product causes injury to the end user, the problem will be magnified and operations will be carefully scrutinised to determine how best to improve the product. If poor documentation is produced and the consumer cannot use it or interpret what the logistician is saying, it is highly unlikely that the product will continue to be purchased. If the parts begin to break down more quickly than anticipated while the product is still under warranty, the products will be returned, providing the basis from which the market share would begin to deteriorate. If after obsolescence the product causes severe waste management or disposal problems, then consumers will have a high reluctance to use it and again this will cause sales to fall. As was indicated earlier, quality cannot be inspected into a product, but must be designed into it. It is extremely important that one faces the quality problem early in all of the work performed, so that quality becomes foremost in everyone's mind. If after obsolesence the product causes severe waste management or disposal problems then consumers will be reluctant to use it and again this will cause sales to fall.

These three elements, schedule management, money management and quality management, are essential to ensure that the ILS management team effectively performs its duties. These lead into the establishment of an ILS skills base. We need to be able to assure ourselves that once we have established the three elements, the skills will be made available to the people required to perform the tasks. The skills lead to successful support systems analysis, developing a support engineering staff, making sure that the production support is in line when it is needed, and developing technical publications so that they can be understood by everyone concerned.

Customer training also becomes a major aspect of the total package of essential ILS skills. If our staff or the customer's personnel are not properly trained anything one may do or say regarding the quality, or the level of performance, or the effectiveness of the equipment one is selling will be worthless, and the position in the marketplace will begin to decrease. This aspect of ILS is supported by the field service group. These are the people responsible for ensuring that once a product has been placed in the consumer's domain, experienced personnel are available to repair it when it fails and to answer any questions that the consumer might have. These queries might concern projected improvements in the product, lack of understanding of the application and use, or how best to purchase replacement parts so that the assistance of the field service personnel is not necessarily required every time the product fails.

These are but some of the items reflected in Fig 2.1. (The remainder of the items will be discussed in later chapters.) What we are attempting to do as part of total ILS management is to ensure that we have consumer approval and a cost effective support package available to the consumer, so that when a product is fielded, the consumer can obtain the necessary parts within a minimal period of time. Concurrent support is essentially another product line. Designing a product to include the capability to be upgraded becomes a necessary aspect of corporate growth. Concurrent support should work in conjunction with concurrent design, the latter being a new phase in the development of a product life cycle. As one begins to field products, the design engineer starts looking at new technologies that could be integrated into the existing product base. This is done in parallel with the production of an existing product. Concurrent design works in such a fashion that, as technology starts to change, the design base keeps up to speed with what is going on in technology and is ready to market an upgraded product early, ensuring that the marketplace is retained. Concurrent support goes along with concurrent design. The logistician should maintain a purview of what is going on in the new design, so that he and his logistics element managers can concurrently look at the product base already fielded, to determine how much of the available material can be made readily accessible to the new designs being conceived. It is essential that the logistician maintain, as with the design of a new product, as many of the support aspects and as much review of what design engineering is doing, so that he is not caught unawares.

However, to ensure that this goal is achievable, we must look at what is required from a management point of view to initiate the work that satisfies this. We believe that a proper work breakdown structure should be provided to each of the lower-level indentures. The work breakdown structure becomes the guidance by which all future tasks are prepared, so that each of the logistics elements described in the following chapters are properly controlled to ensure integration with the other corporate functions.

Fig. 2.2 will assist the logistics manager in defining the interrelationships of functions necessary to ensure proper management control. The figure illustrates a matrix operation where the product is divided into at least five levels of structure. These sublevels define the design of the product down to its lowest plausible level. Commonly most management structures stop at the assembly level, unless a critical component is essential for product operations.

The intersections of these layers are equated to functional management levels. These are divided into management substructures. Each level in the management structure is overlaid with the product work breakdown structure so that cost accounting packages can be developed. This approach permits the establishment of work packages containing the levels that will be performed by each of the engineering and support functions and the

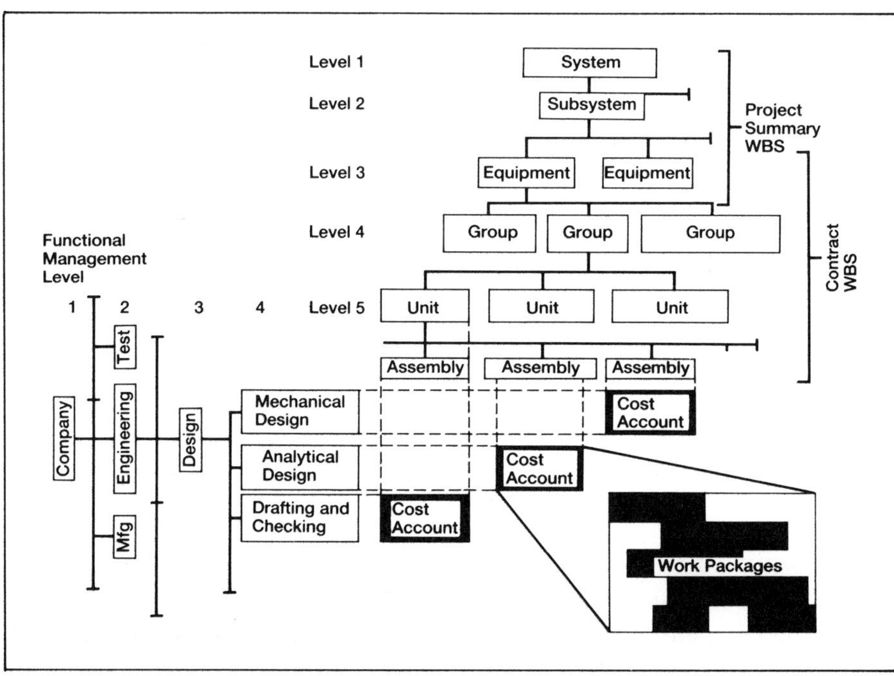

Fig. 2.2 Work breakdown structure (WBS)/functional integration.

amount of funds allocated to perform these tasks. All of these work packages are directly related to the programme milestones. The logistics manager uses this methodology to exercise control over his supporting elements.

Using this approach to programme control, the logistics manager can now initiate his efforts to achieve his ultimate goals:

- Maximising the availability of parts and personnel for consumer acceptance, while
- Minimising the life cycle cost of ownership.

If, in fact, these two goals are achieved, it will mean that we have been able to influence the product and create an effective support system design.

The three basic objectives (influencing the product, creating an effective support system and implementing timely support system requirements) become the very foundation from which a logistician is required to work. In order to accomplish these, the logistician cannot arbitrarily work within his own environment. He must be willing, if not for the sake of the job, then for the sake of profit, to work directly with the design function, because if this is done, he will be able to achieve these objectives.

Working in partnership with the other functions gives the logistician the ability to rapidly assess, and at times assist in, the correction of a specific design problem. The logistician should be an extension of the engineering staff, working with the design engineering group in defining the requirements to achieve the lowest cost of ownership using a minimum number of support people.

All elements of logistics must be tied into the total programme so that as problems arise the element managers can quickly translate them from problems into design solutions. Constant trade-offs must be performed between the design and the logistics efforts so on-time evaluation can be effectively performed. Waiting until the design engineer has completed his analysis would not permit the appropriate trade-offs to be performed.

Logistics planning and management

The keystone to effective logistics support is systematic and orderly management. This should be geared to both logistics support and other related disciplines. Logistics support is flexible enough to permit the professional to tailor his management effort; however, the process also includes sufficient effective inter-relationships of all of the elements to reduce the chances of costly logistics errors through oversight.

Overall management of logistics requires that the professional establish communications that permit the changes in one or more of the elements of logistics to have their full effect on the other logistics elements as well as design. To ensure this, the basic integrated support task must be planned early in the concept formulation phase. Fig. 2.3 illustrates the specific programme management elements required, when they should occur and how and when they overlap. The initial phase that supports integrated support tasks is early management planning, which covers the period from concept formulation through contract definition. This phase of the logistics support management cycle establishes the basic requirements from which all future logistics aspects are generated. This, in essence, describes the ILS specification or requirements document, defining the needs of logistics that will be carried through the total life cycle.

Chapter 1 discussed the generation of the integrated logistics support plan (ILSP). This specification is the starting point from which all of the other logistic support elements are created. The needs, requirements and support functions are in essence defined so that as the product matures, the specific requirements to support ILS can be refined and developed throughout the product's life cycle. Precise, systematic requirements definitions are developed during the early contract definition phase and are achieved through design, development, testing and production. These precise requirements determine when, by whom, why, how and how much logistics will be integrated into the total package.

The efforts that are accomplished during this period will be discussed later in this chapter. What we are looking for at the beginning of the process is the assurance that all of the support elements (specifically maintenance planning, support and test equipment, transportation, packaging, packing

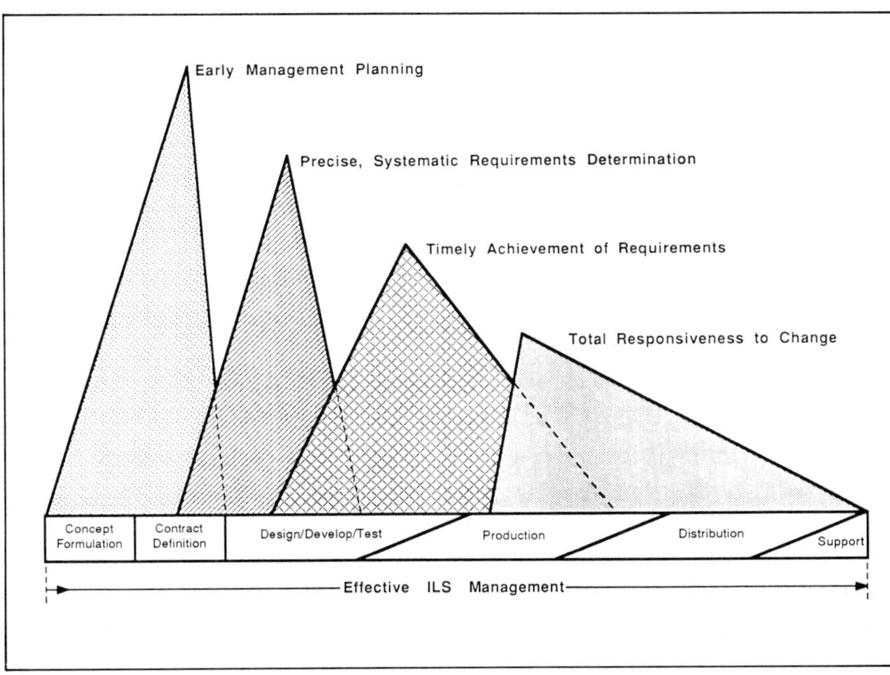

Fig. 2.3 Basic integrated support tasks.

and handling, technical data, facilities, personnel and training, logistics support resources, logistics support management information and environmental impacts) will be appropriately addressed throughout this design phase. Last, but not least, is the total responsiveness to change: configuration management. (Chapter 6 will cover this in much more detail.)

To properly ensure that logistics is in fact integrated early in the design of the product, the logistics manager (LM) must take the initiative, working in conjunction with the programme manager. The LM will prepare a work breakdown structure, which defines the necessary path of the requirements to be performed by each of the logistics element managers (LEMs). This work breakdown structure will be his guidance document, from which all work will be generated. It will be the tasking used to initiate each of the supporting logistics elements that will be performed in conjunction with the system design and product engineering efforts. These inputs, along with the supporting inputs of logistics, will be the basis by which all future efforts are initiated.

It is recognised that logistics and design must be integrated because the inputs from both of these functions become part of the logistics support criteria and objectives base, ultimately leading to the generation and reporting of analyses, the results being the material necessary to support the logistics support analyses database. The work breakdown structure described earlier is prepared so that not only do the logistics element managers know what is required of them, but they also can be properly integrated into the total programme effort. Tasks such as maintenance planning, support systems, analyses, etc., must be performed and managed by the logistics support manager so that he can work cooperatively with the design function in performing the appropriate trade-offs essential to ensure that the contract requirements are achieved. Fig. 2.4 outlines the approach by which the logistics manager utilises the work breakdown structure to initiate the tasks to be performed by his supporting staff.

Balance of trade-off variables

It is extremely important that the logistics manager, along with the other supporting managers, takes the necessary steps to ensure that as trade-offs are performed, all of the engineering and support variables are measured against one another to maintain a proper balance between what we call performance capabilities, modularity and commonality objectives and system availability requirements. This half of the scale is balanced with the cost of ownership and the essential needs.

Before the trade-off variables are described, Fig. 2.5 will be discussed. This balance has been constructed to give the beginner an understanding of what should be traded off in Fig. 2.1. Whilst looking at this from the logistics point of view, it must also be understood that it has a major impact on what

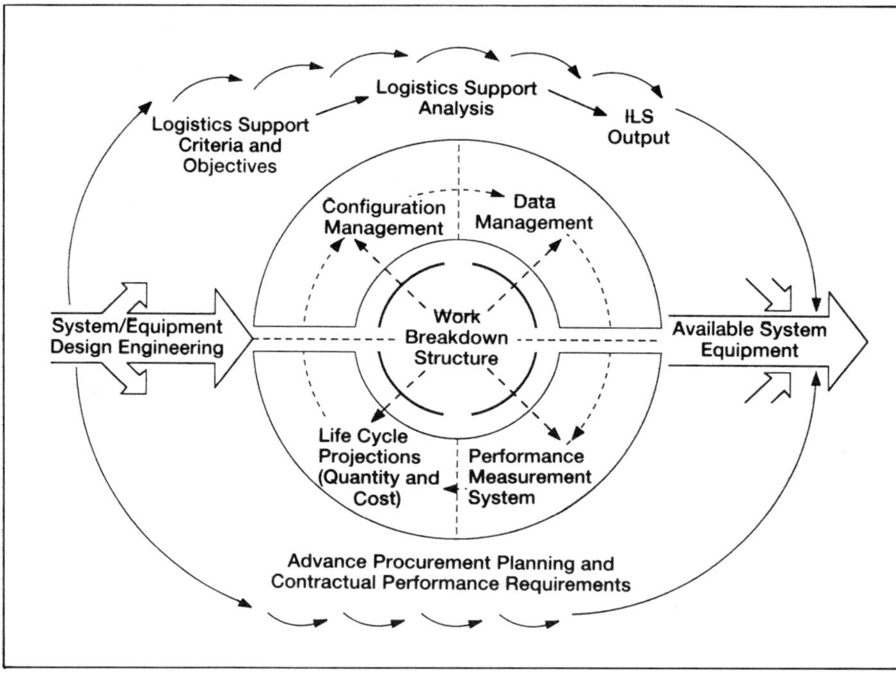

Fig. 2.4 Management initiatives.

goes on in engineering design. The design engineers must assure themselves that the proper evaluations are performed so that when compromises are required, appropriate actions can be taken to maintain the necessary balance among all of the requirements. Starting on the left side of this chart the first elements to be measured and balanced are the functional requirements and system configuration. And the question to be evaluated is, when the requirements stipulated in the design specifications cannot be met, what alternatives must be reviewed to ensure that these characteristics can be achieved? This must be weighed against the system's configuration. How many alternative designs can be defined to show that it is possible to meet the functional requirements? What trade-offs must be made to ensure that the functional requirements can be achieved? What impacts will these trade-offs have against the logistics support functions and how can they be achieved to meet the requirements of reliability, maintainability and support? It must be recognised that the first two items (functional requirements and system configuration) are truly engineering development efforts, but also true logisticians need to recognise that engineers cannot produce the product without ensuring the integration of logistics support. These two elements are then balanced against performance capability. Again, the performance capability is defined by the requirement specification. How they are achieved again must be balanced against the functional requirements and the system configuration.

On the left side of this scale are the elements that are directly related to ILS. First is the installation plan – where will this product be installed and

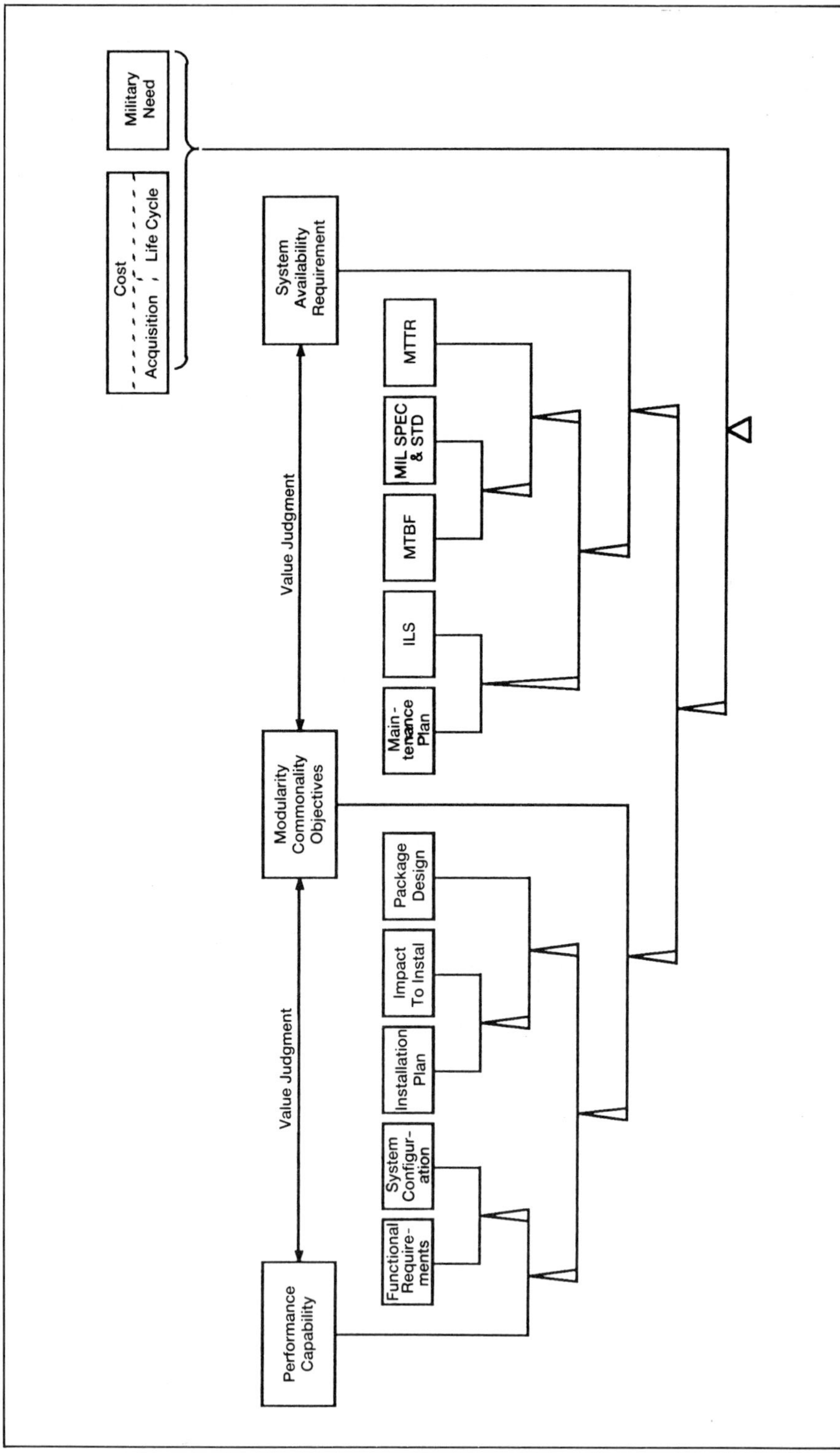

Fig. 2.5 Balance of trade-off variables. MTBF, mean time between failures; MTTR, mean time to repair.

how big is it? How much of an effort will be required to ensure that the product can fit onto the platform for which it is being designed? Can the functions be performed? Are sufficient capabilities built in to ensure that, once installed, the product can be disassembled and removed from the platform without having a major impact on the system? This is balanced against the packaging design. Though it becomes a major activity in full-scale engineering development, we must recognise that the way the product is assembled will have a direct impact on the ILS effort, specifically in the area of maintenance. Performance capabilities are then measured against modularity and commonality objectives. Both of these are true value judgements and must be maintained that way to ensure that all reasonable approaches to design can be measured and achieved.

Logisticians should also recognise that any change to the design must be measured against its total life cycle cost (LCC). Therefore the logistician should ensure that as changes are suggested an LCC is run. This LCC is used to identify what the total cost of ownership will be after the product is fielded. The LCC is not used as a cost accounting approach, but must be integrated into the total design approach and used as one of the management tools necessary to measure optimisation of total support.

The middle portion of the scale represents the efforts between the engineering functions, specifically, the reliability and maintainability (R&M) and the ILS functions. This balance is essential because R&M is the major driver of ILS. R&M is also an integral function of design and if we intend to maintain a balance between ILS, R&M and design, each of these elements must be traded off to ensure the appropriate approach to support. The R&M and logistics functions are then balanced against system availability requirements. Again, that value, a quantitative value, has been defined within the specification. The system availability requirement is then balanced against modularity and commonality objectives, which establishes the basis for the simplification of logistics because it will trade off the number of spare parts required to be fielded with any one design, against the performance capabilities of the product. This balance must be maintained. Trade-offs will be performed, but one cannot outweigh the other.

Each of the functions that will be integrated into the team effort associated with product design must be on a parallel and equal status. The three elements (ILS, R&M and design) are then measured and balanced against the cost of acquisition and the total life cycle cost, as well as the consumer's need. If the scales are not maintained in balance, there is a high probability that the design will not achieve its intended objectives.

Effective ILS management

To ensure that the appropriate information is disseminated among all of the functions, the programme management office must maintain control. Fig. 2.6 represents how programme management acts as an umbrella over the

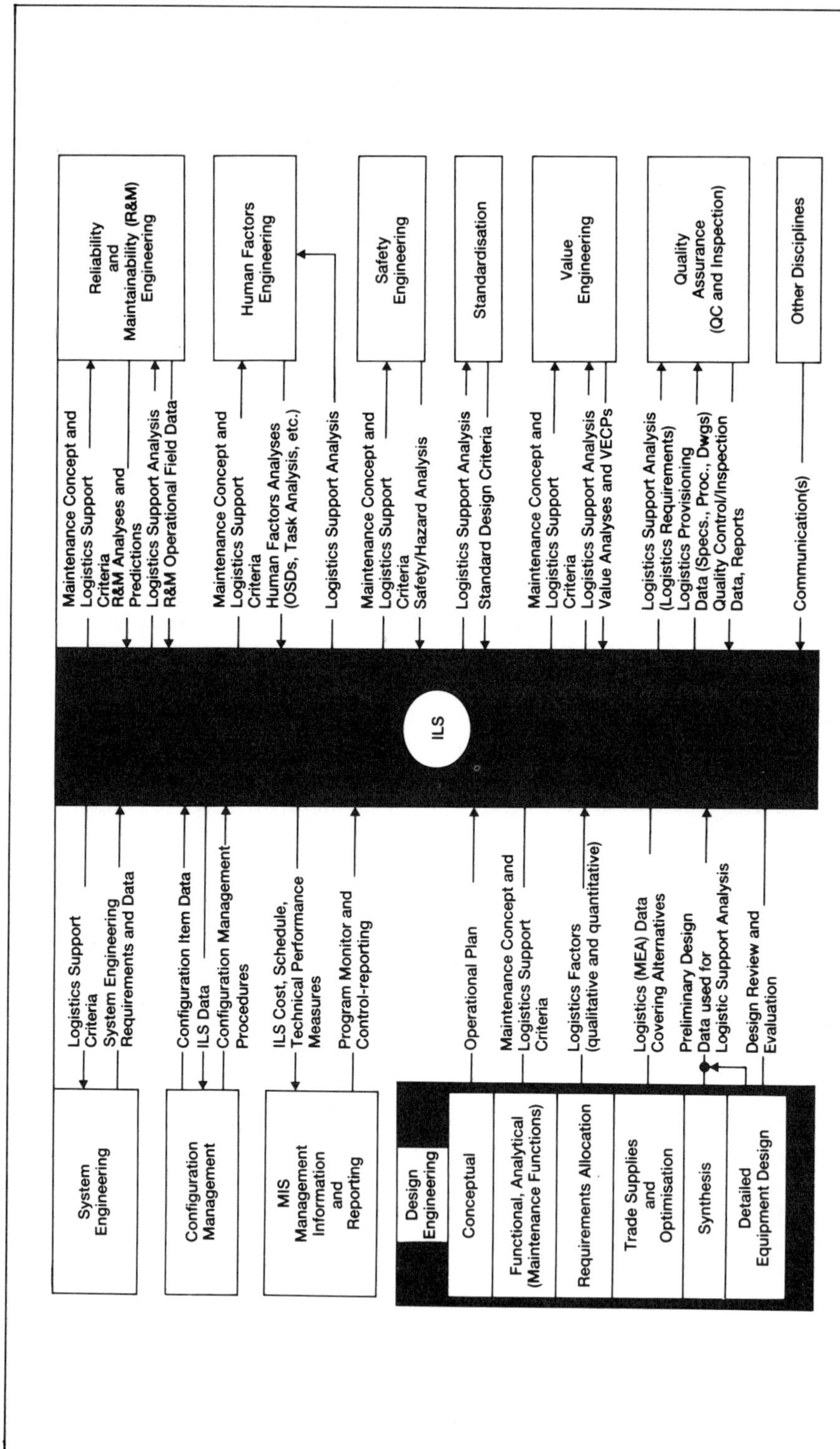

Fig. 2.6 Programme management.

necessary interfaces that are performed and must be maintained within all of the supporting functions to ensure that logistics support is accomplished. ILS is the integrated function among all of the support elements. The information transfer becomes essential if ILS is to work. ILS is highlighted here because the ILS manager must be in a posititon to communicate directly with the programme manager. This role is essential so that the necessary information transfer can be readily accomplished. ILS, therefore, is on the same level as the design engineering elements, quality assurance and other supporting design functions. The information flow between these functions is easily defined, as in the case of design engineering where the ILS elements assist in the generation of specifications. This is to ensure that ILS assumes a major design role when the product is delivered to the design engineer. Configuration management and data exchange are essential elements for ILS, because without them any product that is fielded, if not properly controlled, will in fact create an impossible task for the field or service engineer.

The bottom line is reasonable ILS support and a good profit yield. If essential logistics support techniques and purview of the entire design group is maintained, the logistician can have a positive impact on the corporation's profit line. The logistician's basis therefore is to establish and ensure ILS planning. It is important, though, that one concedes the point that not all of these essential aspects will be accepted and encouraged by management.

The last aspect of management is the planning technique. The logistics manager needs to establish the road map for the industrial base. He should establish the needs and requirements for logistics early on so that they include each of the major support elements that will be defined in the subsequent chapters of this book. This roadmap will outline the typical support planning events for the product's life cycle to ensure that everything is done in a timely manner. Scheduling becomes an aspect that not all of us like to do, but that is essential if we are to meet and accomplish the end item goals: fielding or distributing a product that is acceptable to the customer.

Planning techniques include the best thinking of a representative working group. Putting the team together early and establishing the guidelines by which each member of the team will work, will give not only the team, but also the organisation and the corporation, the visibility necessary to ensure an appropriate and proper marketing base. It provides an approach through which, by the further application of ingenuity on the part of each logistician, one can tailor the specific requirements to the programmes at hand. This gives the ability to take a long and searching look not only at the team, but also at what they are attempting to do, so that all of the positive aspects necessary to ensure that a product is deployed, fielded or distributed to the consumer have been properly analysed, evaluated and traded off so that the various elements of all of the functions have been accepted by the management team. This process will ensure that the deployed product is the best the company can produce.

Appendix A (pp. 223-226) was prepared to assist the professional in determining the requirements to support a fully integrated operation. This appendix outlines the ILS support for a product. Although not all-encompassing it represents the general requirements of each ILS function. It should be used only as guidance. The various applications can be tailored to the specific needs and requirements of the functions planned. Again, this is a guide and should be applied in that way.

The logistics challenge

Because of the constant uphill battle being experienced by logisticians, they are asked to do a lot more than is normally required by many of the other functions within the corporate structure. Fig. 2.7 presents the logistics challenge. It is the ability to accomplish more with far less. It is a task sometimes formidable in nature, but one should face this challenge as all of the other challenges that logisticians encounter. It is not insurmountable, but it is time consuming, and it is an established premise that all must consider.

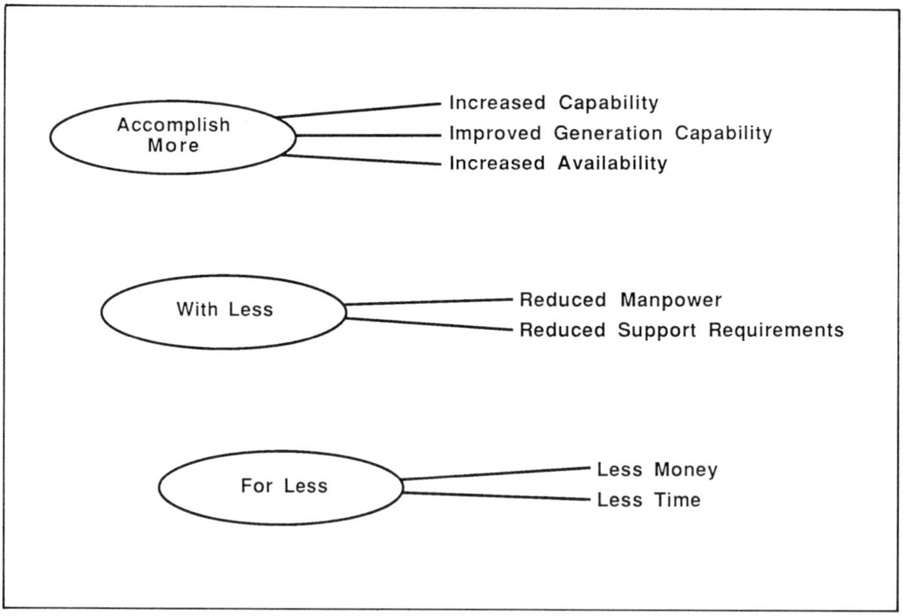

Fig. 2.7 The logistics challenge.

Chapter 3 ACQUISITION MANAGEMENT

This chapter will discuss the first element of the logistics problem – the acquisition process. The first chapter emphasised that during the acquisition process the logistics manager should ensure that the proposed product meets the user's operational requirements. This requires that, from the earliest juncture, support planning is integrated into the entire design and development of the product through integrated logistics support. ILS provides a systematic, orderly approach to the early planning process beginning with identification of the product's need and requirements, then into the acquisition phase, and throughout the system life cycle.

The acquisition process or phase is based on the mission of the product. It focuses on the primary logistics principle of making the right goods and services available at the right place and time. The term 'acquisition' simply means examining the design requirements, assessing the technological essentials, determining the funding requirements, and budgeting, followed by identifying the procurement objectives, methods and details in terms of the whole throughout the product's life cycle. This phase, of course, includes the components necessary to develop, maintain and sustain the product through obsolescence and disposal. Acquisition applies to both the private and public sectors, from the development of a product costing millions to the casual purchases of expendable office supply items. It spans the identification and management of spare parts for a space platform to the operation and supervision of an inventory storage room in a small boutique.

The acquisition phase begins with determining the procurement management objectives and how they contribute to the effectiveness of the product. The procurement and contracting functions are subsets of acquisition, as are statutory, regulatory and company policy requirements, including tax implications. The acquisition process uses models and strategies to verify optimal management techniques, environmental changes or implications, the stability of requirements, interface potential with multiple users, the impact of constrained resources and employment of innovative system concepts.

As for funding, the acquisition phase budget process should include, but not be limited to, initial cost estimates, inflation effects, single-year versus multi-year budgeting and procurement cost advantages, service flexibility and the interconnectivity of functions. Risk assessment also has to be accomplished during this phase to determine the estimated effects of uncertainty.

Finally, the acquisition phase deals with contracting and negotiation, and international procurement effects and decisions.

From a technological perspective, the process should include the use of state-of-the-art or high technology increased costs, as well as the acquisition considerations involved in whether the system and its components are technologically sound.

Public and private sector acquisition

The acquisition phase is essentially identical in the public and private sectors. The goals, objectives and methods are practically the same. The differences between the two, however, may be noted by identifying these peculiarities of the public sector (government acquisition). Briefly, these are:

- Public funds are used and generally expended only as prescribed by law. The government sector has rigid budgetary and accountability restrictions, and public auditing procedures are required.
- Items acquired are normally for the use of several requisitioning agencies, bureaus and departments. Acquired items are not intended for resale.
- Employees in the public sector are public servants who may be merit system employees.
- Except for governmental clandestine operations, the acquisition process is conducted in full public view. In fact, the United States has an extensive system of 'Sunshine Laws', which requires complete openness to ensure that the acquisition operation is a matter of public record. Public sector acquisition requires accountability to the public at large, not the shareholders, stockholders or a sole proprietor, as in the private sector.

- Acquisition is a prescribed process in the public sector with rigid legal and administrative restraints. In many countries, even the parliament or the congress is actively involved in the process by imposing constraints.
- Managers in the acquisition process are subject to censure by the parliament or congress, the public at large and the media, such as television or the press.
- Finally, governments act in a sovereign capacity in the acquisition process. Government has no competition against itself, but it does force vendors to compete with one another, and it usually selects and accepts the lowest bid or price, often at the expense of quality, durability and even effectiveness.[1]

Private sector commercial characteristics differ widely from those listed for the public sector. In this case, the 'bottom line' or governing criterion is normally competition and profit.

The key to understanding the differences between public and private acquisition lies primarily in the objectives of each. In the private sector the goal is profit and improvement of the firm's competitive advantage over other firms. In the public sector the objectives are multiple and unclear. Some view economic efficiency as primal. Success is measured subjectively. For a more extensive coverage of the differences between the private and public sector acquisition processes, the reader's attention is directed to Stanley Sherman's *Government Procurement Management (1985).*

The acquisition process logistics functions

One of the components of the acquisition process is materials management. Harry Page, in *Public Purchasing and Materials Management (1980)*, sees a close relationship between purchasing as part of the acquisition phase and materials management. He states that materials management should serve as an integrator of the flow and control of materials and services. It begins with identifying needs and terminates (according to Page) with product delivery to the client or customer. Actually, most life cycle-oriented logisticians differ with Page to the extent that materials management terminates much later, after obsolescence and upon completion of the waste management and disposal processes. Page does not consider materials management as synonymous with logistics management. Though we disagree with his concept on the extent of materials management, we can use his premise for the functions of materials management to more aptly review the logistics functions, and modify them somewhat to gain a clearer

---

[1] Sherman, Stanley (1985) *Government Procurement Management*, (Gaithersburg, MD: Wordcrafters Publications), 2nd edn, pp. 6–7.

perspective of the acquisition process logistics functions.[2] These are outlined as follows:

- Coordinating and determining needs. This function includes analysing product demands, documenting design requirements, determining inventory status, estimating parts availability (commonly referred to as 'estimating pipeline status') and conducting status control and quality assurance measures.
- Generating requisitions, sometimes referred to as 'requirements determination', and conducting status control.
- Preparing the solicitation or announcement, called the 'request for proposal', to request bids or proposals, and implementing status control and quality measures.
- Proposal and contract negotiation.
- Issuance of the purchase order or contract.
- Order acceptance with notification that includes delivery status to the production department. Traffic control and warehouse coordination should also be effected.
- Receiving the items or parts, including preparation of the receiving reports. Again, quality assurance and acceptance measures should be implemented.
- Inspecting and accepting of delivered items.
- Implementing warehousing and distribution operations.
- Concurrently implementing document assembly, verification and payment procedures.
- Developing 'vendor purchase and item history' file.
- Marketing.
- Conducting acquisition phase coordination and follow-up with the production operations department.
- Conducting physical distribution and transportation functions, to include shipping, insurance and export-related document preparation.
- Forecasting requirements to maintain specified levels of customer service and support.
- Conducting documentation, training, and mobility arrangements to ensure timely, reliable maintenance and operational capabilities.
- Implementing sustainability actions is then necessary to support life cycle requirements. Sustainability implies the capability to provide the user or customer with the required parts or items to keep the product operational or to restore operational capability (if necessary) throughout the system life cycle. The absence of attention on sustainability functions is one of the principal failures of manufacturers that result in customer disenchantment with the system, loss of good will and emergent customer loss of loyalty to the brand name.
- Implementing disposal actions, to include reviewing the inventory for excess or obsolete items, re-distributing or selling these items as re-

[2] Page, Harry R. (1980) *Public Purchasing and Materials Management* (Lexington, MA: Lexington Books), pp. 21–6.

quired, and planning for ultimate disposal of the system when its life cycle is terminated. This function necessitates planning and management for obsolescence, and should include planning for effective waste management.

A special note is important pertaining to the last point. Practically no manufacturer or government agency considers the importance of disposal planning – planning and management for obsolescence and planning for effective waste management – as part of the acquisition phase process. Effective logisticians with a life cycle orientation know this too well and are currently trying to educate top management in the necessity to follow through with the entire life cycle process in the acquisition phase *before* systems are developed and fielded. The popularity of short range planning, requirements for immediate return on investment, the maximum profit motive and management's prevalent 'not on my watch – it's not my problem' syndrome preclude most top managers from even fantasising about disposal actions let alone accommodating these requirements to the acquisitions process.

The nature of acquisition management in DOD

The most sophisticated, orderly acquisition process in either the private or public sectors is practised by the US Department of Defense (DOD). For this reason this manner of acquisition management is discussed more fully here. The US military services implement strict methods and procedures because of costly extensive weapons systems that require quality, reliability, readiness, complexity and logistics support. The sophistication of these products in an era of tighter budgetary constraints presents a challenge unparalleled in acquisition management. In actuality, although there have been some recent problems in executing the acquisition process, the military services techniques are exacting to the extent that other government agencies have implemented similar procedures, and some civilian industries and manufacturers emulate the DOD system.

The Department of Defense acquisition system

The authority for DOD acquisition is vested in Office of Management and Budget (OMB) Circular A-109, *Major System Acquisitions*, dated 5 April 1976. This circular, the DOD 5000 series of instructions and the military services policy statements and regulations define the phases and procedures applicable to each segment of a major system acquisition. DOD Directive 5000.1, *Major and Non-major Acquisition Programs* (1 September, 1987), prescribes the acquisition policy for use by the military services.

Major acquisitions are initiated by the military services based on a perceived need which is then translated into an operational requirement. For example, the US Navy calls this a Tentative Operational Requirement or TOR. This requirement was issued to develop an options document that describes the range of potential options available to satisfy the need. The choices are reviewed, and a formal document is issued which specifies the exact capability and characteristics desired for the system. This document then becomes a part of the military service's Programme Objectives Memorandum (POM), an annual document prepared by the military services and submitted to the Secretary of Defense by April every two years. However, the military services conduct a mid-year POM review during the off year. POMs recommend the resource requirements and programmes for the military services and must be consistent with the SECDEF's fiscal guidance. The POM is a major document of the PPBS.

Many other steps are required but have been omitted for the sake of brevity.

In each military service, most weapon systems go through at least four general phases which are discussed below.

1. The *concept exploration phase*. This begins with the service's issuance of a programme charter which designates a programme manager, commonly referred to as the PM. Now finalised, the programme is included in the President's budget that is submitted to the Congress in January of each year. The Congress then *authorises* the requirement and *appropriates* funds. Meanwhile the PM ensures that the conceptual studies are conducted on such issues as new R&D technology applications, production quantities and the use of existing systems.

The proposed system is divided into intermittent design approval phases, each of which is preceded by a milestone. Thus the milestone for the concept exploration/definition phase is milestone 0, which establishes the requirement for the programme, approves its initiation, and provides the authority to budget for it. The following are considered:

- Mission area analysis.
- Affordability.
- Life cycle costs.

In addition it is determined whether modifying an existing US or Allied system could provide the desired capability and an operational utility assessment is done.

As the system approaches milestone 0, the military service and SECDEF decision, the following items are prepared:

- A *concept paper* which includes an analysis of the alternatives developed from the results of the conceptual studies and an analysis of attendant life cycle cost trade-offs. A final recommendation is included.

- A *test and evaluation master plan* describing all of the required tests for evaluation. The plan must conform to the prescriptions of DOD Directive 5000.3, *Test and Evaluation*, and other military service regulations.
- *Defense system acquisition review council (DSARC)* briefing materials that include the total results of the Concept Exploration phase results for the DSARC formal review. After this action, a SECDEF Decision Memorandum is prepared that terminates or continues the programme.

2. *The demonstration and validation phase.* This consists of the PM's results of the system's Statement of Work (SOW), the solicitation and other documents that may be necessary to contract for demonstration and validation.

The decision here is milestone I, the concept demonstration/validation decision. Milestone I considers seven essential elements:

- System alternative trade-offs.
- Performance, cost and schedule trade-offs.
- Appropriateness of the acquisition strategy.
- Prototypes for the system and its components.
- Affordability and life cycle costs.
- Potential common-use solutions.
- Cooperative development opportunities.[3]

The results of the milestone I decision are a broad programme cost, system schedule, and operational effectiveness and suitability goals and thresholds. These provide the programme manager with a great deal of flexibility to come up with innovative, cost-effective solutions.

As a rule, contracts are awarded to commercial contractors for at least two of the optimal alternatives. The contracts provide for adequate demonstration models to evaluate the preliminary design. These contracts also require the submission of development specifications and results of independent operational test and evaluation conducted by the contractor. Consistent with the essentials of life cycle management, the PM also ensures that the following plans are completed:

- Programme Management Plan.
- Integrated Logistics Support Plan (ILSP).
- Configuration Management Plan.
- Test and Evaluation Master Plan.

The programme is now ready for the Milestone II SECDEF decision, a major decision for continuing the programme. Milestone II requires the submission of the Decision Coordinating Paper (DCP) and the Integrated

---

[3] Commander William C. Keller, *The Defense Resource Allocation Process* (Newport, RI, Naval War College, June 1988 (Rev.)), p. V-13.

Program Summary (IPS) which has been prepared in accordance with *DOD Instruction 5000.2, Major System Acquisition Procedures*. The culmination of the Milestone II decision provides the basis for a decision to place the system into full-scale production when the Full-scale Development phase is accomplished.

For milestone II, full-scale development decision, the Defense Acquisition Board considers 13 areas:

- System cost affordability in terms of the military value of operational suitability and effectiveness.
- System programme risk in terms of the benefit of the added military capability.
- Planning factors for the transition period from development to production.
- Industry surge and mobilisation capacity in realistic terms.
- System stability programme impact factors.
- Potential common-use solutions.
- Prototype and demonstration/validation results.
- Milestone authorisation.
- Assessments of manpower, personnel, training and safety requirements.
- Procurement strategy in terms of programme costs and risk assessments.
- Plans for integrated logistics support.
- Affordability and life cycle costs.
- Command, control, communications and intelligence requirements to include communication security.[4]

3. The *full-scale development phase*. This begins following the SECDEF approval in Milestone II with the award of a contract. The PM normally awards the contract to one of the civilian demonstration and validation contractors to design, test and document the system. The contractor is expected to complete the work consistent with the Demonstration and Evaluation Phase programme plans. The result of this phase should be a fully designed and tested system with a technical data package to provide for a competitive production by other contractors.

4. The *production and deployment phase*. This begins with the Service Secretary's approval to produce the product. It consists of the PM's issuance of production contracts that also include the Programme Management Plan, the Integrated Logistics Support Plan (ILSP) and the Configuration Management Plan. As products are produced, they are deployed by the military services. As a rule user units are selected for follow-on operational test and evaluation (FOT&E). Test results are analysed to ensure full operational

[4] *Ibid.*, p. V-13.

capability in terms of operational requirements, and to determine if product improvements are indicated.

Milestone IV, logistics readiness and support review, reviews the assurance of operational readiness of the product and its support objectives for the first several years of use. This review normally occurs two years after the product is initially deployed.

Acquisition categories

The centralised systems acquisition procedures described above apply primarily to costly major programmes. Some DOD smaller acquisitions fall into less centralised categories. In these cases the military services alone can make the determination for systems acquisition, and all have prescribed procedures for doing so. The Navy uses acquisition categories (ACATs) that provide for lower-level decision making and less stringent requirements for documentation in acquisition management. The Navy has four ACAT categories:

1. *ACAT I.* These are systems designated by the SECDEF which meet any of the following criteria:

- Systems with significant risk in development, urgently needed and those with specific SECDEF interest.
- Systems jointly acquired by two or more military services or Government agencies.
- Systems with acquisition costs exceeding $200 for RDT&E or $1 billion in procurement funds.
- Systems with high congressional interest.

2. *ACAT II.* Such systems have two components:

- ACAT IIS systems have the Secretary of the Navy as decision maker, and pertain to systems that have any of the following: $100 million in RDT&E or $500 million for procurement costs; high congressional interest; significant technical, cost or scheduled problems; mission criticality; or unusual demands.
- ACAT IIC systems require Chief of Naval Operations (CNO) approval, and apply to those with similar exceptions as for IIS.

3. *ACAT III.* These systems require milestone decisions by the Deputy CNO or Directors of Major Staff Offices, and apply to systems that affect military characteristics or the Navy's combat capability, and do not exceed the dollar thresholds above. The ultimate decision maker is the Vice Chief of Naval Operations (VCNO).

4. *ACAT IV*. These are programmes that do not fall into the ACAT III category which require Operational Test and Evaluation. Again, the final decision maker is VCNO.

Recent initiatives to improve DOD's acquisition process

DOD acquisition, because of its large expenditures of the federal budget, has become the favourite target for Members of Congress and the media. DOD has recently taken several initiatives to improve the acquisition process, some of the more significant of which are outlined below:

- DSARC reviews have been simplified to better depict the impact of weapon systems outyear costs.
- Cautious implementation of the use of multi-year contracts by the military services to optimise acquisition costs with economical production runs.
- Intensive management of Preplanned Product Improvement (P^3I) programmes to encourage system changes in smaller increments and reduce risks.
- Emphasis on the implementation of the provision of the Competition in Contracting Act (CICA) to increase competitive procurement to reduce acquisition costs.
- Encouragement of the military services to implement initiatives to reduce acquisition costs, and the costs for maintenance and sustainability through the use of ILS.

The role of integrated logistics support (ILS) in DOD acquisition

The total requirements and logistics support for DOD systems necessitate the adequate application of the ILS principles. Logistics support must be *integrated* to support the system from the earliest portion of conceptual development. Fortunately, as described in detail in later chapters of this book, DOD insisted that ILS be considered as a disciplined approach to:

- integrate support requirements into system and equipment design;
- relate support requirements to design;
- develop and acquire the required support throughout the system life cycle.

The DOD policy, DOD Directive 5000.39, requires integrated logistics system support, and pertains to planning and resource decisions, ILS management requirements and management support requirements. Acquisition programmes must have an ILS programme from the earliest onset.

Contracting considerations in the acquisition process

There is not enough space within the scope of this book to detail all of the contracting considerations relevant to acquisition. Most of the contracting provisions are contained in the Federal Acquisition Regulations (FAR). However, there are some general guidelines that we provide as a guide to contracting with the Government that will assist in logistics.

Solicitations for dealing with the Government are advertised in the *Commerce Business Daily* (CBD) at least 30 days before contract negotiation. Solicitations may fall into three categories:

1. Open competition relates to CBD notices of the availability of a Request for Proposal (RFP) that requests firms to prepare a proposal and bid on the needed work. Because of all the machinations in handling this contracting process it may take from 6 to 12 months to get a contract.
2. Selected competition is used when the Government agency determines that just a few firms are qualified to do the work and solicitations are encouraged from them. Normally these firms have 30 days to respond, and the time-frame is about 3 to 7 months to go to contract.
3. Sole source contracts are based on the conclusion by the Government that only one source can perform the work because of technical and other reasons. Government contracting officials have become so concerned about criticism and potential disciplinary action, that a sole source contract is most difficult to obtain.

There are seven reasons, all specified in CICA, why a Government agency may opt to use the sole source, non-competitive procedure for contract award:

1. There is no other responsible source other than one firm who has the product or service that can respond to the Government's needs.
2. An unusual or compelling urgency would be seriously injured unless a specific firm provided the product or service.
3. The award of a contract to a particular firm is necessary to maintain that firm as a source in the event of mobilisation or industrial mobilisation; or if an educational or non-profit institution requires the award to enable an essential engineering research or development capability to be established or maintained.
4. If the source is restricted by an international agreement or treaty, or if a foreign government reimburses the procurement costs to the US Government.
5. Brand name commercial items for Government use.
6. For national security reasons to prevent the disclosure of an executive agency's requirement.
7. When, in terms of the public interest, the head of an agency determines it is better to use non-competitive contracting. However, in this case 30 days' prior notification to Congress is required.

Contract types also fall into seven categories:

1. *Firm-fixed price contracts* are those whose price is not subject to adjustments following award. Most of these contracts are for existing off-the-shelf items. There can be some pre-stated modifying provisions, however, such as allowances for inflation, labour costs or materials costs.
2. *Cost-reimbursement contracts,* as the name implies, provide payment for allowable incurred costs as specified in the contract. *Cost-plus-fixed-fee* contracts fall into this category and include a negotiated profit fixed at the start of the contract.
3. *Incentive contracts* tie the contractor's profit to delivery of the services or technical performance. *Fixed-price with incentives* and *cost-reimbursement with incentives* fall into this category.
4. *Quality contracts* have three subtypes. *Definite-quantity contracts* call for the delivery of a definite quantity for a fixed period; *requirements contracts* have the same provisions, except that delivery is only on Government call; and *indefinite-quantity contracts* provide for an indefinite quantity within a stated limit for a specified period. Funds for items or services are obligated by each delivery order, not the contract itself.
5. *Time-and-materials, labour hour and letter contracts* are used for work in terms of specified rates for direct labour costs per hour, and normally contain price ceilings.
6. *Agreements* are written understandings that provide for subsequent contracts for delivery. They are NOT contracts. The most common is a *basic-order agreement* (BOA) which provides the terms for future contracts or orders, a supplies or services description and methods for pricing, issuing and delivery of orders in the future. BOAs are often used when the Government cannot specify items or services in advance.
7. *Purchase orders* are used to obtain specific goods or services that are well defined and under a certain price, $10,000 for Government agencies, except DOD which is $25,000.

Obtaining bids for items or services is based on Government issue of Request for Quotes (RFQ) which comes in two forms, (1) *Invitations for Bids (IFQ)* which request quotes from contractors with the award to the lowest bidder; and (2) *Request for Proposal (RFP)* which provides a general description of the work, and asks firms to indicate how they will do the work and what it will cost. Again, the lowest bidder is normally selected in terms of capability and approach.

Proposals

Proposals respond to RFPs by firms to choose to compete for the work. The Government prescribes no particular format, but convention in practice dictates a comprehensive approach. Normally, a proposal contains the following sections:

- Section I: Introduction, overview and executive summary.
- Section II: Problem understanding, a section in which the firm indicates its knowledge and understanding of the problems raised in the RFP.
- Section III: Technical approach, where the proposer describes all the tasks it will use to accomplish the work and produce the materials or services.
- Section IV: Corporate experience, that lists the previous contracts that relate to the work required.
- Section V: Corporate and project management, a section describing the firm's internal organisation and how it plans to manage the work, to include experience and qualifications of its personnel who will be associated with the contract.
- Section VI: Facilities and security, that describes the firm's physical layout, technical support, services and status of security clearances (facility and personnel).
- Section VII: Cost proposal, a section usually separated from the rest of the document for objective reasons, that describes what the expected price will be.

Logistics support analysis (LSA) and acquisition management

LSA is an in-depth analysis that uses a systems approach to evaluate the impacts of trade-offs in the design, support and operational characteristics of the system. LSA provides a capability to:

- Evaluate the alternatives for optimal hardware configuration and logistics support.
- Identify logistics and manpower requirements.
- Provide a record capability, called the LSA Record or LSAR.

The ILS manager directs the establishment of the LSA, and it is normally accomplished by a developer or manufacturer. It should begin before the earliest phases of acquisition management, normally at the conceptual development stage. A more detailed discussion of the LSA and the LSAR may be found in Chapter 7 pertaining to ILS.

Acquisition innovations in industry

Some enterprising firms have taken the lead in introducing more advanced techniques and automation for managing the acquisition process. For instance, Clay Beskin wrote an article entitled 'Is JIT spelled barcoding?', which appeared in the October 1986 issue of *Manufacturing Systems*. Beskin wrote that portable data collection and scanner laser technology not only provide more timely collection of information, but provide flexible application with increased reliability for improved inventory control, materials handling and tracking, and acquisition functions.[5]

An article in *Production Engineering* in March 1986 reported that Scientific Computer Systems, Inc. replaced their manual material procurement system with an E-Z-MRP microcomputer system from C.R. Smolin, Inc. The new automated system provided SCS with control over its material requirements and procurement process, and detected potential problem areas in advance of their occurrence. The main benefits from their perspective were improved order tracking and acquisition continuity.[6]

Scott Garske and Kathleen Murray reported in the May 1988 issue of *Hospital Material Management Quarterly* that there is increased evidence of life cycle equipment asset control in the health care industry. An audit trail and control can now be provided for each piece of hospital equipment for planning, budgeting, purchasing, receiving, tagging, tracking, control, depreciation and disposal. A bar code labelling system is the cornerstone for system control in this asset control system.[7]

These examples indicate a slow, emerging trend in the private sector to automate and sophisticate the acquisition management process. Much more research, development and application should be done to integrate the acquisition phase more fully into life cycle logistics management and ILS.

Conclusion

This chapter examined the nature of the acquisition process from a total life cycle logistics perspective. Acquisition identifies the procurement objectives, methods and details in terms of the whole system by means of the ILS concept. Today's challenges for product excellence, quality and sustainability require that a logistics capability is integral to the system's acquisition. This requires highly trained and skilled logisticians to direct and manage the procurement process.

It demonstrated how the acquisition phase fits into the total ILS framework by meeting procurement objectives. Acquisition differs in the public

---

[5] Beskin, Clay, 'Is JIT Spelled BARCODING?'. *Manufacturing Systems*, **4** (10), October 1986, pp. 62–64.

[6] Anon., *Production Engineering*, **33** (3), March 1986, pp. 68–70.

[7] Garske, Scott and Murray, Kathleen, *Hospital Material Management Quarterly*, **9** (4), May 1988, pp. 35–40.

and private sectors, but the former has a more detailed, sophisticated process. Acquisition in industries and manufacturers tends not to be as stringent, except for the defence industries. Although there may be differences, it is suggested that private industries could profit greatly by adapting many of the principles imposed by the Government. The logistics nature of acquisition can be best demonstrated by an examination of the relationship between purchasing and materials management, and how both tie into the integrated logistics system.

Chapter 4 ESSENTIALS OF PRODUCTION OPERATIONS MANAGEMENT

The reader need only review the classified ads section of most Sunday newspapers to see the endless litany of job positions in production operations management. Most of the job descriptions call for an appropriate degree and experience in engineering or previous experience in the production field. What rarely is called for is a logistician. Reviewing the curricula of many business schools a subject called Production Operations Management is normally a required course, but it is usually taught by a faculty member in the Management Department, Marketing Department or Operations Department. This person usually has a doctorate in business administration or operations research, and has little, if any, idea of how production operations really work in the field. Such a person is rarely a logistician.

Both of these examples demonstrate the prevalence of a misconception: logisticians too frequently are not involved in the production operations field. Then we wonder why the manufacturing and services sectors perform so poorly.

Life cycle logisticians provide the solution: these professionals consider production as an integral part of the total flow of materials from concept development and design; through the acquisition process; continuing through production, delivery and use by the customer; and terminating in obsolescence and waste management. Aware of the system life cycle, logisticians contribute to availability factors (the probability that the system will work and be ready when needed for use), reliability and sustainability. In other words, the logisticians' value is their emphasis on *all* aspects of the system, product or service life cycle. This is particularly important in today's resource constrained environments.

In its most rudimentary form, production operations management is the systematic, orderly direction and control of all of the processes that convert inputs into finished goods and services. Inputs consist of workers and managers, capital (assets, funds, equipment and facilities), material, utilities and information. Types of conversion vary and include the following types of information:

- Physical (the place of business).
- Locational (a highway).
- Attitudinal (a TV drama).
- Educational (a college).
- Physiological (a health clinic).
- Informational (a computer program).
- Exchange (a market).
- Storage (a warehouse).

Key to these are the customer and information feedback loops with information from external sources, economic trends, markets and other material sources.

Manufacturing and services operations contrast

Manufactured products are material and normally are durable, although some manufacturers design-in obsolescence. Services are intangible and expendable, and sometimes perishable, like concepts, ideas and data. Manufacturers often provide services such as credit, warranty or repairs.

Goods are inventoried, stored and transported, while services are not pre-produced. There often are no inventories to soften unpredicted customer demands on the production system. Services also have high degrees of customer contact, while most users of goods have no contact with the producer. Instead, customer contact is provided by retailers or distributors. Some service firm employees have low customer contact, while others have high customer contact.

Another differentiation is demand response time. Manufacturers have long lead times, while services with direct customer contact must be more responsive.

A final contrast is in measuring quality, which is much easier with tangible goods. The services industry is more vulnerable to individual preferences in assessing quality.

The world of production operations

Production operations is one of the functions in an organisation, like accounting, finance, marketing, human resources management and engineering. It is also a profession, such as production control manager, or operations manager. Production operations is also a system of decisions, such as problem recognition, problem formulation, analysis of alternatives, choice and implementation of the preferred alternative and assessment and evaluation of the selected decision.

For optimal effectiveness, many firms employ a logistician who is a well-trained generalist to guide the organisational operations and transform inputs into outputs through disciplined productivity, quality and cost effectiveness. The logistician is ideally suited for this function because he is well versed in many disciplines such as quantitative methods, organisational behaviour, manufacturing and production, operations, management information systems, business economics, international business, business ethics and law.

The field of production operations management covers the following areas in producing products and services:

- Product plans, especially what goods and services should be offered.
- Competitive priorities in terms of quality and cost.
- Positioning strategy of the product or services.
- Management quality.
- Process design.
- Logistics support in terms of production.*
- Technological considerations and choices.
- Capacity planning and location.
- Facility layouts.
- ILS of the product or service.*
- Maintenance and sustainability.
- Master production scheduling.
- Production and staffing programmes.
- Materials management in terms of supplies.
- Inventory control and management.
- Priority scheduling.
- Quality control management.
- User feedback analyses and support.*
- Waste management.*

For a detailed explanation of the above, the reader's attention is directed to any of the better production operations texts, among which are Heizer and Render, *Production Operations Management: Strategies and Tactics*. However, almost none cover the items designated in the list above by an asterisk (*). These fundamental areas are critical for production and services firms, and ignoring these issues is a primary reason why good firms and their products and services can fail. It is in production logistics support, ILS which includes Logistics Support Analysis (LSA), user feedback analysis, support (customer satisfaction) and waste management that the life cycle logistician more than earns his or her keep.

Process design

Simply stated, process design involves the blend of people, machines and work requirements (often referred to as transformations) that cohesively will provide the product or service. Process design is the key to production operations, and includes determining the selection of inputs, operations, logistics support requirements, work structure and methods to produce the product or service. It is the key because of:

- technological changes that severely modify the product or service;
- changing competitive priorities in the firm;
- demands that result in volume changes of the product or service;
- effects of competitors with similar products;
- changing costs or availability of inputs such as raw materials or personnel technical expertise;
- results of user feedback analysis;
- the firm's continued financial ability to maintain market leadership.

Process design decisions have strategic importance. They are closely linked with product design because wrong choices can affect the firm's competitive position in the market. Ignoring logistics support costs can also hamper optimum profits.

Most writers who have addressed production operations have indicated that process design has four basic elements:

1. *Capital intensity*. The greater the relative costs of equipment and human skills required to operate it, the greater the requirements for capital.
2. *Resource flexibility*. The degrees to which equipment and employees can deal with alternative products, functions and requirements.
3. *Vertical integration*. The process that products or services undergo from raw materials through customer use. High vertical integration implies that this chain of events occurs primarily within the company's own control.
4. *Consumer involvement*. The degree to which the customer is integral to the production operations process.

Production planning and control

Production planning and control are integral to production operations management, as they ensure that product or service supply meets product or service demand.[1] Customer demand forecast is based on sales projection periods. This demand is considered an *independent* variable, since it is not under the control of the consumer. By way of contrast, a *dependent* variable occurs when one component is required as an element to complete the finished product. Once consumer demand is predicted, the supply of the finished product is determined from the supply of available inventory or from producing the product. Production planning involves determining the number of units to be produced, times and intervals to produce them and required availabilities of materials and machines to produce the necessary units.

This process requires consumate materials planning and control, again coordinating supply with demand of materials. Here the forecast of raw materials is dependent on the production schedule. The process, called materials requirements planning (MRP), provides for the flow of materials. Most production facilities consider this a closed loop system that begins with determining material requirements for production and ends with passing orders to vendors and receiving their goods. Treating this process as a closed loop system is one of the fundamental fallacies of any manufacturing system, because it isolates itself from contingencies that may dissipate the purposes of the process, such as vendor supply problems, labour problems, quality assurance issues and the like. To ameliorate situations like these, many manufacturers have become enchanted with techniques such as a 'just-in-time' (JIT) production system.

JIT production is a technique innovated by the Japanese and hailed as one of the reasons for their high degree of success in competition with other producers. It is considered a 'new way' of managing manufacturing. JIT is based on two fundamental tenets: (1) inventory's primary purpose is to serve as a problem cover-up; and (2) inventories are a waste of time and money.[2] Recognising the goal of eliminating inventories of raw material and work-in-progress, the results of JIT are lowered operating costs, increased operations efficiency, improved product quality, elimination of carrying costs for finished inventory and reduction of disruptions in production. Increasingly, a number of manufacturing firms are spending a great deal of money attempting to emulate Japanese manufacturing. But so far, JIT has had little impact on the services industry.

Although it has become popular, the direct application of Japanese forms of JIT has resulted in operational problems in the United States. Reasons

[1] Coyle, John J., Bardi, Edward J. and Langley, C. John Jr (1988) *The Management of Business Logistics* (St Paul, MN: West Publishing Co.), p. 137.

[2] Ibid., p. 140.

why this Japanese method of production operations management has run into difficulty are as follows:

- The US geographical dispersion of suppliers and vendors, unlike Japan, makes JIT access too difficult. For example, Ford Motor Company's recent suggestions to its vendors that they relocate to the Detroit, Michigan area were not well received.
- The US's inadequate transportation infrastructure is not conducive to JIT. Rail systems are inadequate (or non-existent) to the extent that sole reliance on the more costly trucking industry is necessary.
- US traditional socio-cultural habits, attitudes and economic system emphasise independence and competition among vendors, unlike Japan.
- The American business operational psyche mandates short-term profits, contrary to fiscal goals and objectives of firms in other countries. JIT requires a systems approach to financial planning that has continuity over years.

American management focuses on JIT from two perspectives: (1) to reduce carrying costs associated with inventory storage and handling; and (2) to avoid production disruption costs due to forecast inaccuracies because of excesses or shortages in finished goods inventory.

Materials management

As a continuation of production operations management, the logistician is concerned with materials management. The key here is a coordinated systems approach for optimal results. The logistician's work here is particularly appropriate. He ensures that the purchasing department's primary objective of minimising the purchasing cost of items through volume price discount buying is balanced with the inventory control department's objective of minimising inventory carrying costs through low inventory levels and small quantity purchases. Therefore, a carefully orchestrated, integrated approach is provided by a professional who accomplishes both objectives while meeting production operations requirements.

Effective materials management is interdisciplinary and covers the activities relating to production planning, scheduling, purchasing, shipping, storage, handling and control of all materials necessary for the manufacturing and services operations processes. Materials management entails determining production requirements, production process scheduling and procurement, and procuring, storing and dispensing materials at minimum cost. It is critical for efficient production operations management.

Materials management from a logistics perspective should be adopted for the following reasons:[3]

---

[3] Ibid., pp. 129–36.

- Its use provides a capability to use the latest control tools for effective decision making, such as operations research (OR), electronic programming and profit centres.
- It achieves cost reductions by coordinating purchasing, transportation and inventory reduction.
- It provides a capability to reconcile competing objectives by demonstrating the effects of economic or political trade-offs.
- It provides coordination and control of materials subfunctions to monitor work flow levels and advance information, both of which enable systems to function efficiently.
- It demonstrates the consequences of 'make and hold' decisions where suppliers now carry finished goods inventories previously held by manufacturers.
- It centralises the materials authority to support production operations management more effectively by improving coordination.

Materials management, on the other hand, also has limitations. Among these are:

- The reluctance of manufacturers and the services industry to find and hire logisticians who are experienced in systems analysis and also have expertise in materials management.
- The reluctance of various departments, such as personnel, purchasing, physical distribution and transportation and finance to coordinate with one another.
- The belief that production control is too important to be subordinated, and conversely, that purchasing cannot be subordinated either. In other words, there is an emerging trend of 'fiefdoms' of specialities in organisations.
- Unwillingness to incur the costs of effective materials management.
- Combinations of all of the above, which result in lower productivity, loss of quality and decreased profit margins.

For an excellent discussion of materials management, the reader's attention is directed to Garry J. Zenz, *Purchasing and the Management of Materials*.

Inventory control

Inventory is critical to manufacturing and services firms, although in terms of a percentage of gross national product (GNP) inventory has been declining in recent years due primarily to deregulation and information management. Coyle, Bardi and Langley (1988) state that inventory represented about 20% of the GNP in 1974, and by 1986 it was slightly over

14%.[4] Nonetheless, inventory is an asset to the firm, represents a percentage of value added and provides for enhanced customer satisfaction. On the other hand, due to trends for more product lines, inventory levels and costs are escalating.

Inventories provide many advantages, as they provide for:

- purchase economies;
- transportation savings;
- production capabilities;
- enhanced customer satisfaction.

Safety stock

Most writers in the area of production operations management consider safety stock in the management decision-making process as a type of insurance to provide inventory when stock falls short of forecast demands. Life cycle logisticians consider this a fallacious, stop-gap, poor management practice. Thinking of safety stock inventory as a type of 'bank account' against which to draw stock is poor management. Safety stock should never be relied upon because of stockouts. These stocks should be set aside only for use in emergencies or contingencies beyond the control of the manufacturer or service industry provider.

It is difficult for management to determine how much of a buffer, or safety stock, to have on hand. Too much means excess inventory. Generally accepted practice requires determining the percentage of carrying costs, which includes capital cost, carrying cost, storage cost, inventory service cost and inventory risk cost. Then this percentage is multiplied by the monetary value per unit and the number of units involved. An overriding factor is that safety stock levels require the use of probability analysis, otherwise carrying costs can be too high. Yet the cost of lost sales due to stockouts can be higher. Of course, the more recent management practice of causing intentional stockouts to increase demands, and raise profit margins is unethical, unsatisfactory management, and considered a total disregard for customer service.

A final point should be made concerning inventory control. The '80/20' notion, or Pareto Principle, is a management decision based on marketing experiences that found that 20% of the buyers make up for 80% of the sales, or beliefs that 20% of the people do 80% of the work. We have found that this principle seems generally true, but as we approach the last decade of the twentieth century, the Pareto Principle should be modified to indicate '90/10' phenomenon. Because of high unemployment, extremely low productivity and poor worker attitudes toward work, it appears to us that 10%

[4] Ibid., pp.157–9.

of the people get 90% of the work done. Real estate brokers indicate that 10% of their agents bring in 90% of the sales. Proprietors indicate that 10% of their customers account for 90% of their sales.

Other production operations functions

Production operations management is concerned with other facets of the production process, such as internal traffic, receiving and inspection, quality assurance, waste management and transportation scheduling.

Internal traffic pertains to managing inbound traffic of materials. It requires close coordination with the transportation and physical distribution department, which handles modes of transportation, arrival times and optimal costs reduction. Unfortunately, many vendors control this function instead of manufacturers. This situation can be avoided by alert, conscientious buyer management and effective procurement and contracting.

Receiving and inspection focus on ensuring that the goods that are received are the goods that were ordered, and that the items are in good condition. The system to handle discrepancies must be effective, to include prompt notification to production operations managers, the purchasing department and the finance disbursing office. Expedited or 'rush' shipments should be avoided at all costs, as premium payments are required.

Quality assurance is an active inspection and reporting effort to make sure that inbound materials as well as production items and services meet all tests for quality. Discrepancies are handled by notifying the purchasing office, which notifies carriers and vendors, the physical distribution and transportation department and the insurance department. Much has been written about the low quality standards of American producers. The American penchant for inspecting a limited statistically determined sample of items has become ineffective, and serves as a motivating factor for adopting Japanese methods of production and quality assurance.

Waste management is concerned with the disposal of salvage, scrap, excess and obsolete materials. This is a much-neglected function in the United States because of its disposal societal tendencies. Some scrap materials cannot be sold and have to be disposed of safely and in accordance with Federal regulations and common sense. One can simply refer to the daily newspaper for numerous accounts relating to improper waste disposal, particularly toxic waste.

Chapter 5 RELIABILITY AND MAINTAINABILITY

Integrated logistics support is not intended to stand as a functional island. It must be integrated with all of the other functions that together produce a successful product. A logistician would recognise early that two engineering properties have a direct impact on the success of support: reliability and maintainability (R&M). R&M are systems engineering functions that are included as requirements in the product definition, commonly called the system specification. There is a dichotomy between these two functions. Reliability is truly engineering in nature and is the most significant function influencing support, since it provides a measure to determine frequencies of maintenance actions. Maintainability, though, is the means with which the defective item can be restored to operability.

Regardless of how reliable a piece of equipment may be, if the module that fails cannot be removed, the product cannot be maintained. Because of this, maintainability is traditionally an engineering function, even though the work that is performed as a result of it is a logistics function. The three elements, reliability, maintainability and maintenance, when working in conjunction with one another, have a direct impact on every logistics action that eventually takes place. A change in one or more of these elements has a devastating effect on all of the logistics support.

One must recognise that this needs to be a very tightly organised, closed loop system because if what has happened in the past is not measured, the future cannot be projected. That point has been emphasised to re-establish

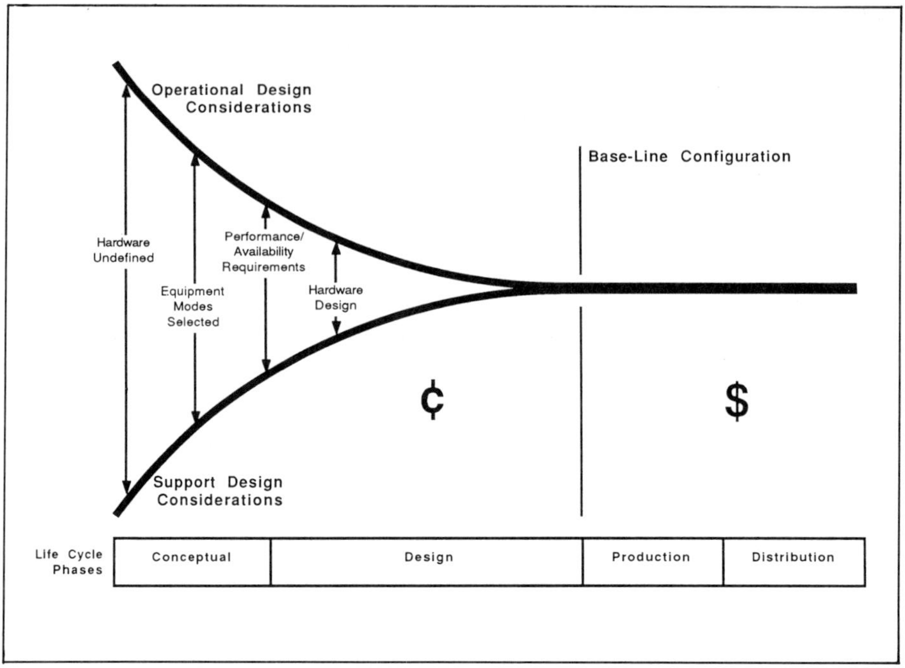

Fig. 5.1 Support impact on design.

our reliability goal; to not include it directly in the design will have a major impact on what the maintenance analysis profile will be. Maintenance analysis, maintenance concepts or maintenance planning (see Section 7.1), regardless of what it is called, will drive each of the logistics functions. Logisticians must constantly be aware of its impact and make sure that the established goals are realistic, not because they want to say 'this is the best thing since the invention of apple pie', but because they want to do this from a realistic stand and, as a result of their inputs in the total functional picture, to be able to maintain the balance in functional requirements as defined in Chapter 1. This can be measured from the point of view of when and how these elements are integrated. If logisticians spend their time properly at the very beginning of the programme, the cost to the programme will be in pennies. If it is allowed to proceed beyond the early portions of the design phase, it will cost much more to integrate the proper R&M into the design (see Fig. 5.1).

Reliability

Reliability is one of the most significant functions influencing support since it provides a measure for determining the frequency of maintenance actions. From the point of view of modern engineering technology, reliability is characterised as:

- The probability of performance over a required period of time.
- The comparison of available strength against probable stress.
- The trade-off of reliability against other desired qualities.
- The cost required to reach a given reliability goal.
- The volume production of the reliability inherent in the design.
- The optimum utilisation of a product after delivery.

Therefore, reliability could be defined in the following terms: reliability consists of the performance characteristics of a product that reflect its ability to operate satisfactorily in order to complete the product's assigned function. A more precise definition of reliability is the probability of successful operations for a specific period of time under specific conditions and environments of operation.

This definition contains the term 'probability', which indicates the use of a quantitative measure. In addition, the definition of reliability contains other considerations: successful operation, period of time and conditions of operations. The successful operation of a product is generally defined in terms of its intended use. For example, an automobile will operate satisfactorily when driven on the highway, but will not operate as well off the road. From the definition of successful operation, the definition of failure can be derived. A failure is the inability of a product to successfully perform its intended function (we might also add to safely perform its intended function). For the above definition of reliability to be meaningful, there must be an understanding of what constitutes successful performance in order to continue use of the product.[1]

Before proceeding to a discussion of reliability, the question that is normally raised is how much reliability is economical? High reliability requires increasing the cost of research and development of materials, manufacturing and testing. Against these increases is the cost of field service, troubleshooting, spares, repairs and maintenance. As reliability goes up, maintenance operations and maintenance requirements go down. Fig. 5.2 presents two sets of cost curves, which are provided in relation to the levels of reliability. The sum of the two represents total cost.

It therefore becomes obvious that the greater the reliability provided at the onset of a programme, the less likelihood there is of having a major increase in support functions. Reliability engineering therefore provides a collective and dedicated approach to design and production, causing reliability to be built into the equipment through prediction methods, controls, measurements and reports, and later on in the design through analysis of failure phenomena.

In other words, the reliability engineer's objective is to place failure control on a quantitative basis and to plan for it to achieve the level of failure rate that represents maximum cost effectiveness. It is therefore important to

---

[1] Constantinides, A. (1986). *Basic Reliability* (Consolidated Lecture Notes, Tutorial session, 1986 Annual Reliability & Maintainability Symposium), PBR 2.

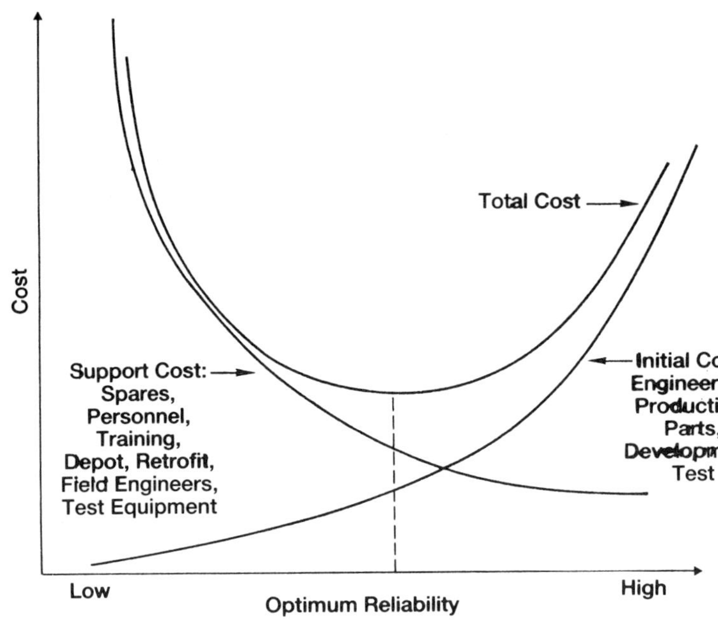

Fig. 5.2 Reliability versus total system cost.

modify the traditional disciplines and recognise reliability in the trade-offs between performance, schedule and cost. It is again reflected in the balance of trade-offs necessary to ensure that all of the functions in Fig. 2.5 maintain the programme requirements and cost.

To further define reliability, the period of time the product is to operate is a more definitive base to which we can apply some mathematical values. A mission's (or function's) duration is defined as covering some specific interval of time or some specific number of cycles of operation. Once the criteria for successful operation have been established, the operation of the product can be compared to the criteria for the required time periods. These criteria for successful performance may change with time.

The last part of the definition of reliability is the stated conditions of operation. A product is designed to operate in a given manner under particular sets of conditions. These include environmental conditions, such as temperature, pressure, humidity, acceleration, vibration and shock. Also included are operational conditions such as voltage, current, torque and corrosive atmosphere. Both environmental and operational conditions will be encountered in manufacturing, transportation, storage and use. If a product fails when operated in its intended environment, it is not satisfactory. However, if the product is subjected to stress in excess of its design, the failure may not be a reasonable measure of unreliability. In summary, reliability is a performance characteristic that reflects the ability of a product to operate successfully for a long enough duration to complete its assigned mission.[2] Reliability is a quantitative measure defined in terms of probability.

Reliability is expressed in terms of mean time beween failures (MTBF); from this attribute, the probability of completing one or more specific profiles can be calculated. When the mission profile is a definitive requirement, probability of mission success (P_s) can be used rather than MTBF. In normal cases, either P_s or MTBF, not both, is specified for any given requirement.

To assess which value should be used, a relationship between the product reliability design philosophy and the overall mission must be closely reviewed. The following discussion is directed toward the design of reliability into products, not to the work of correcting the mistakes that other parts of the hardware design may introduce into a design.

As distinct from equipment design, product design is concerned with organisation and communication as they relate to the design of individual items. In the design of large-scale systems, there is a need to think in terms of the whole in addition to the operation of individual pieces of equipment. Complexity, which characterises large-scale products, is at the root of the need. Complex products may perform many functions, process many inputs and translate and display many outputs. Therefore, only a broad perspective will permit a search for optimum operation.

The product's reliability goal, which is determined by some pertinent measure of system effectiveness, stems from the system concept. Preliminary system design determines the type and minimum number of equipment items in the network; the required configuration of these items to achieve the reliability goal is then determined. After a configuration is reached, an allocation of failure and repair rates is made for each equipment item consistent with the system reliability goal. During the system development, process, continued adjustments and re-evaluations to ensure the achievement of the reliability goal at minimum cost are made.

The overall design activity begins with the system concept and culminates with a set of specifications that are meaningful enough to permit sound planning and comprehensive enough to present a perspective of the product.

Reliability activities

The logistician must fully appreciate the efforts of the reliability engineer. It is necessary for the logistician to work in conjunction with him so that the quantitative and qualitative value, as established for the product, can be

---

[2] Ibid., PBR 3.

properly established with realistic design guidelines and goals. The following set of activities is used by the reliability engineer in the performance of his task. It must be recognised early that reliability is a systems engineering function, working in cooperation with the design engineer. The criteria established by the reliability engineer become the basis by which the design engineer performs his duties.

In today's environment, many new tools that can integrate the reliability function are becoming available to the design engineer. It is not intended that this function be eliminated, but the reliability effort should become more integrated into the design engineering functions so that the reliability role will become more of an audit function rather than integrating itself into the total.

Until such time as reliability efforts become automated, the reliability engineer must perform the activities outlined in Fig. 5.3. Selectively time phasing each of these activities, as well as those in the logistics manager's design function, should be paramount in the reliability engineer's mind.

This book is not intended to go into the mathematical analysis necessary to calculate reliability, and the statistical explanation of the reliability values that ultimately have an impact on logistics. Rather, it is to give the logistics manager the necessary tools to understand what is required of reliability, in order for him to function in consonance with the reliability engineer to support that element as it applies to logistics. The intent is to keep this book at a management level rather than entering into the scientific determinations of establishing reliability values. Therefore, the next series of sections will describe the types of necessary programmatic tasks to be performed in accordance with the programme schedule, and the efforts that have been proposed as part of its activities, in sufficient detail so that there

- Selection and application of components
- Reliability analysis (block diagram)
- Failure mode, effect, and criticality analysis (FMECA)
- Reliability prediction
- Critical useful life analysis
- Design review
- Reliability test and evaluation

Fig. 5.3 Reliability activities.

is a much more satisfactory understanding of what reliability and maintainability will need to do to complete the job satisfactorily.

Before going into the specific programmatic tasks, however, reliability's role in the design needs to be defined. Table 5.1 represents the various activities that the reliability engineer must perform to ensure that the specific tasks and the organising functions are dealt with cooperatively as the design progresses through its life cycle. It represents the authors' cumulative experience in designing the requirements that the reliability engineer must accomplish during the development of any single piece of equipment. One must recognise that not all of these elements must be performed; the ones that are required will depend on the maturity of the design in question (whether it is a new research and development project or something that has been defined as off-the-shelf and is being modified to satisfy a new requirement).

As in logistics, the reliability role is also to minimise requirements and tailor needs to the specific programme in question. If the tendency is to over-estimate the amount of work necessary, the reliability and design engineers may never accomplish what they set out to do, so the reliability engineer, the logistician and the design engineer must work cooperatively in defining the requirements at hand. Table 5.1 provides an understanding of what should be done in the total programme base rather than trying to selectively define the absolute needs for any one design.

Programmatic tasks

To be able to accomplish the activities that are about to be defined, the logistics engineer, along with the reliability engineer, must have a full appreciation of which tasks are required during the life cycle of a programme. To simplify the reliability engineer's approach and the logistics manager's understanding, the programmatic tasks defined in the US Government's MIL-STD-785 have been summarised in Table 5.2.[3] Rather than attempting to define and copy all of the tasks that have been identified in this document, a matrix has been prepared to give the logistics manager some insight as to the reliability engineer's responsibilities.

Table 5.2 divides up the various elements necessary to perform reliability programme tasks. It is structured into four columns. The first two columns represent requirements that define the various programme elements; the results of the defined output that will either lead to the design criteria, design engin-eering needs or to the LSA. The third column is the application of maintenance planning and analysis process, which assists the maintainability engineer and the logistician in establishing the mechanisms, the

[3]US Department of Defense, *Reliability Program for Systems and Equipment Development and Production*, MIL-STD 785.

Table 5.1 The role of reliability ($\bar{R}$) in design

Task	*Responsible organisation*	*Reliability engineering role*			
		Provide	*Review*	*Support*	*Coordinate*
Hardware design	Engineering	• $\bar{R}$ design guidelines • $\bar{R}$ requirements/ allocations • Part derating requirements • PPL	• Designs for – Parts selection – Worst case – Parts application – Failure impact – Compliance with $\bar{R}$ requirements	• Use of 217B. to determine failure rates • $\bar{R}$ risk assessment	• $\bar{R}$ analyses – Stress – Worst case – $\bar{R}$ K-factors • $\bar{R}$ model use – Determine redundancy requirements – Proper $\bar{R}$ choices based on LCC
Maintainability design	Engineering	• FMECA • Failure rates for system	• BIT/BITE for false alarm potential	Trade-offs to select optimum $\bar{R}$ and $\bar{M}$ combinations • BIT/BITE applications Trade-off decisions	• FMECA – maintainability information
Software design	Engineering	• $\bar{R}$ design guidelines • Standards to maximise $\bar{R}$ of software	• Logic for FMECA	Software design requirements	• Modelling of software faults
Quality assurance	Quality	• $\bar{R}$ test requirements	• Failure data • Test results from incoming inspection, parts screening, manufacturing screening and ATPs	• Procedure development – Incoming inspection – Parts screening – Manufacturing screening – Test procedures	• QA process
Supplier management	Contracts	• $\bar{R}$ programme • Quantitative $\bar{R}$ requirements for hardware/software	• Supplier – Failure data – $\bar{R}$ planning – $\bar{R}$ programmes	• Contract negotiations on $\bar{R}$ terms/conditions	• Supplier implementation of MHC $\bar{R}$ programme

Table 5.1 (Continued)

Task	*Responsible organisation*	*Reliability engineering role*			
		Provide	*Review*	*Support*	*Coordinate*
Safety design and analysis	Engineering	• FMEA data for hazards analysis • $\bar{R}$ design guidelines	• Hazards analysis for impact on $\bar{R}$ design	• Trade-off studies	• Safety and $\bar{R}$ design
Human factors	Engineering	• $\bar{R}$ design guidelines	• Human factors design for any impact	• Trade-off studies	• Man/machine interface requirements to ensure adequate failure prevention/ detection/ correction
Logistics support analysis	Logistics	• $\bar{R}$ design guidelines • FMECA • Failure rates 51 R data for LSA	• Maintenance plan for potential failure inducing actions • Planned packaging/ transportation	• Maintenance planning and analysis • Analysis of maintenance/ handling impacts on reliability	• LSA and $\bar{R}$ design for impact
System test	Reliability engineering	• $\bar{R}$ test • $\bar{R}$ test plans • $\bar{R}$ test criteria	• Test procedures • Test results	• Monitor testing	• Monitor CA for $\bar{R}$ impact
Parts and materials control	Reliability engineering	• $\bar{R}$ inputs to PPL	• Standard parts list • Standard materials list • ESD control requirements	• Parts and materials selection • ESD control	• Parts use and trade-offs

Table 5.2 Programmatic tasks

Reliability Programme Tasks			
Requirement	*Results*	*Application to maintenance planning and analysis process*	*Application to design process*
Reliability Programme Plan	Integrate R tasks into R programme	Establishes R programme tasks to estimate replacement frequencies	Provide task requirements for achieving reliability by design
Programme Reviews	Ensure R programme is progressing according to plan	Ensure rel. data information is supplied to maintenance actions	Ensure reliability criteria are being implemented through design process
FRACAS	Collect R data for use in FSD	Analyse potential failures to reduce maintenance actions	Provide design information to improve reliability
Modelling	Apportion R estimates to sublevels	Provide R requirement values to maintenance levels for use in LSAR	Provide reliability requirements to design for all levels (parts, modules, subsystems)
Allocations	Allocate quantitative R requirements to lower levels	Allocate R requirements to level to which maintenance is performed	Allocate R requirements to levels for design engineers
R Estimates	Provide a measure of the R of design	Provide a measure of inherent reliability of each part to use in planning	Used in the iterative design process to assess the reliability of the design
FMECA	Eliminate design weaknesses	Trade off R and M to achieve Ao, scope maintenance, and support programmes	Provide design criteria to increase design strength to decrease potential failures
Sneak Circuit Analysis	Eliminate latent/unwanted safe functions	Identify potentially unsafe anomalies in critical items	Provide priority in elimination or bypassing of unwanted unsafe functions

Table 5.2 (continued)

Reliability Programme Tasks			
Requirement	*Results*	*Application to maintenance planning and analysis process*	*Application to design process*
Tolerance Analysis	Achieve functional performance with tolerance degradation with time	Identify failures as related to operational time	Increase functional failure-free time through design
Part Programme	Provide selection system for standard and non-standard parts	Standardise parts and materials – reduce support costs	Provide design criteria to design engineers on parts and materials approved process
R Critical items	Reduce design and support risks	Identify items requiring special maintenance and support considerations	Identify special design considerations and areas of risk
Effects of Testing, Storage, Maintenance, etc, on reliability	Maintain inherent reliability	Prevent degradation of inherent reliability due to maintenance	Provide design criteria on degradation rates due to maintenance and support
Environmental Stress Screening	Develop plan for FSD	Establish standards for spares; ensure minimum variance of parts in operation	Ensure minimum variance of parts used in design; reduce tolerance build-up in design
Rel. Dev./Growth Test Programme	Resolve reliability weaknesses	Eliminate reliability problems in maintenance and support	Identify areas of design requiring redesign
Prod. R. Accept Test	Verify inherent R in each system	Maintenance and support requirements equal in each system	Indicate R characteristic inherent in each system
Software Reliability	Total hardware/software reliability	Provide for software maintenance	Design criteria for integrated hardware/software reliability

approaches and the appropriate tools necessary to support a maintenance programme. The fourth column consists of the application to the design process and is the integration of the results into the design. These tasks represent a culmination of all reliability programme elements that are needed in a full system. Specific tailoring must be performed as indicated earlier to be within the guidelines of the specific programme in question. The tasks should be carefully reviewed so that a selective number will be included.

System reliability design philosophy

The reliability engineer, as part of his overall goals and objectives, establishes a set of guidelines to be used to define the types of components necessary to meet the required reliability goals. Components such as integrated circuits (ICs), resistors and capacitors are the mainstay of all design. These components have been tested and are defined by the various military functions within the USA. The range includes high reliability products that can survive temperatures in excess of 125°C. They also have been tested so that they meet other environmental conditions. Each of the components screened for, say, Class A can support and work in environments such as a space shot or in missile launching. If the parts fail this category, the manufacturer tests them for Class B screening, which is not as stringent as Class A. It is then determined how many components satisfy that; those failing Class B fall into Class C, and so on. Those failing to satisfy the requirements of Class C could eventually become commercial parts utilised as part of everyday radio and television components. The reliability engineer then establishes the lists of components that should be used as part of the design, based upon the requirements of the programme.

Prior to establishing the application of components, the reliability engineer must familiarise himself with the requirements of the contract so that he can appropriately define the various selected components.

Reliability analysis

The first objective of reliability design is to determine what the probability of mission success/MTBF should be for the proposed equipment under review. The best way to achieve this is through a thorough analysis of the relationship of reliability with the other important parameters of the system; for example, in a radar system, the probability of detection of false alarm rates. This entails a measure of reliability effectiveness that can be quantitatively manipulated to describe the consequences of alternative system design. This also entails developing a definition of failure that allows for reliability estimates. Relating the level of reliability to be achieved to the cost of product design is also important. Within this broad spectrum of system reliability analysis, a set of specifications can be written. These are the bases that initiate the detailed design in a way that will culminate in an optimised system approach.

The reliability specification requirements for function or mission success should include the following:

- A definition of system performance such that every condition is defined as acceptable (success) or unacceptable (failure).
- A definition of the environmental conditions that prevail on the design parts of the product. In addition, consideration must be given to the duty cycle and periods of operation.
- A definition of mission time is a carefully considered quantitative statement of time during which the product must function. In highly complex products, which operate in different functional modes at different stages of the mission or which use certain sub-segments only if conditions require, the time requirements for each sub-segment should be established. If they cannot, it may be necessary to determine probabilities of successful functioning during a range of mission time.

These are by no means the only requirements, but they are the major considerations.

Combining all of the above, a reliability design procedure can be developed:

- Define product reliability in terms of operational requirements.
- Develop an index of system reliability effectiveness.
- Apply mathematical techniques to evaluate the reliability and cost of alternative system configurations.
- Evaluate the consequences of each alternative configuration.
- Compare allocated and predicted values to determine the next best course of action.
- Compare the results of the product requirement and maintenance philosophy against probable adverse impacts on the associated interfaces.

This procedure is by no means rigid. What is important is the systemisation of objectives and the use of analytical techniques. Figs. 5.4 and 5.5 have been included to assist the logistician in understanding the work that needs to be accomplished by the reliability engineer to ensure that reliability is properly integrated into the design.

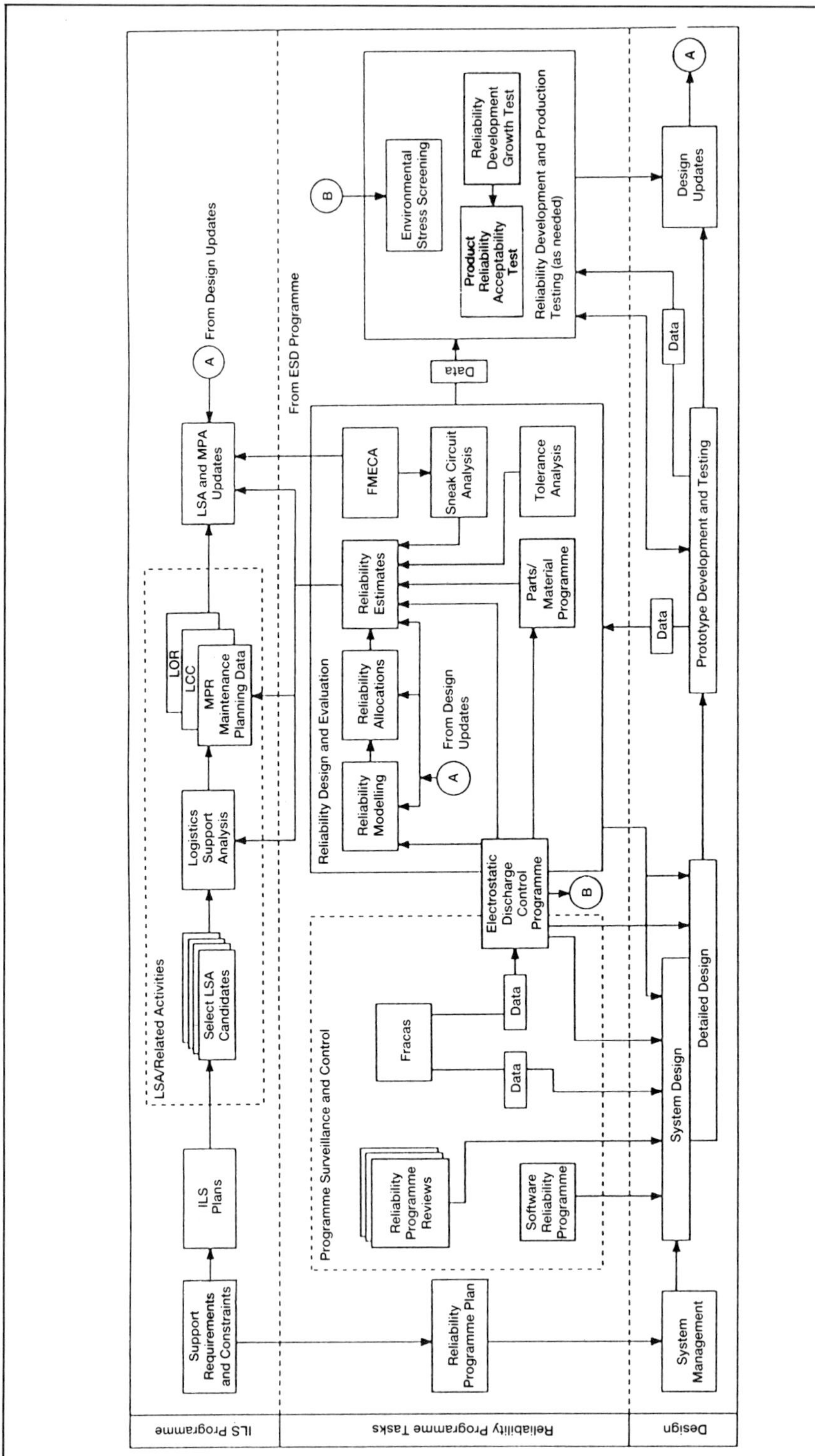

Fig. 5.4 Reliability analysis. ESD, electrostatic discharge; FMECA, failure modes effects and criticality analysis; LCC, life cycle cost; LOR, level of repair; LSA, logistics support analysis; MPA, maintenance planning analysis.

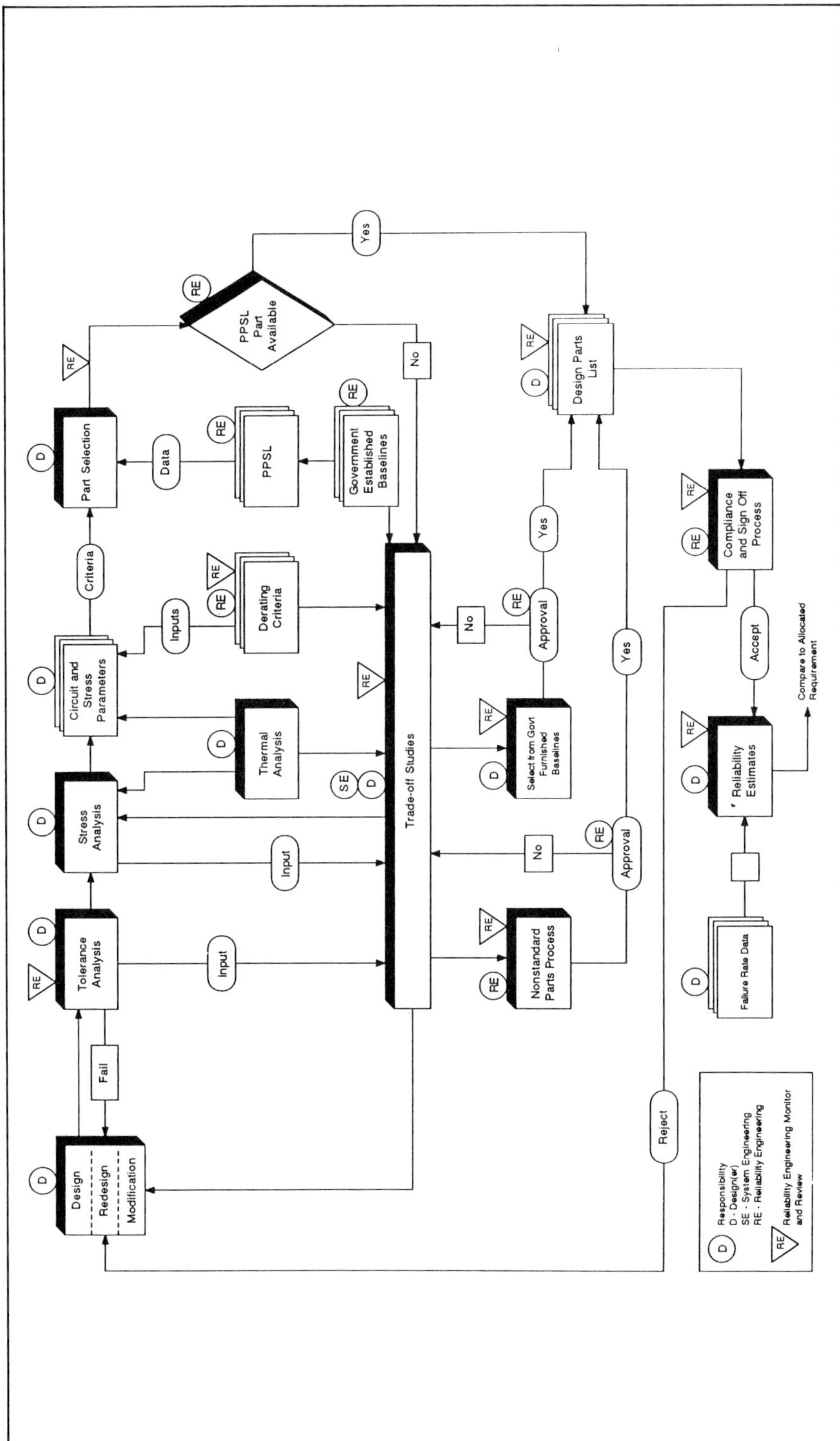

Fig. 5.5 Reliability analysis flow during design phase. PPSL, preferred parts selection list.

Failure modes effects and criticality analysis

It can be argued that the failure modes and critical effects analysis (FMECA) can be performed by the reliability engineer or the maintainability engineer. For our purposes, it has been decided that the FMECA becomes a part of both. It will be discussed here as part of the reliability aspects; it will be further discussed in the Maintainability section of this chapter. It is important that FMECA is recognised as the tool by which the reliability engineer measures the functional impact of a failure on the product. It is the measure of failure to the lowest component, whether it be a resistor, capacitor or IC, or an assembly that has a direct impact on operation of other functions within the design. Fig. 5.6 illustrates the impact of both the reliability engineer and the maintainability engineer on the design. The maintainability engineer should be fully aware of what this means, because it has a direct impact on the type of spare parts required to help support the design after the product has been distributed.

Critical analysis is exactly what it says: the evaluation of the design to the lowest component. It measures the impact of a failure on the product. The inference is that one single component, though it may not appear to be critical when it fails, could actually make the product inoperable.

Reliability estimates/reliability predictions

To be of value, estimates must be timely. However, the earlier an estimate is needed, the more difficulties there will be. It is certainly true that the earlier an estimate has to be made about an unknown future event, the more difficult it is to make a meaningful estimate. As an example, the reliability of electronic equipment is known with certainty only after the product is worn out and its failure history has been faithfully recorded. But a great deal of knowledge about reliability can be accumulated over a short, early period in the product's useful life. Even though the confidence in the estimate is lower, there is some opportunity for it to be influenced. Similarly, considering the stages back through installation, shipment, test, production, test design, development, procurement, etc., less and less can be known with certainty about reliability. However, what is known or estimated becomes a more and more valuable basis for taking action. After all, there is no value in simply knowing that a certain failure will occur at some specific time in the future. The value comes in having the opportunity to do something to prevent the failure from occurring.

The two trends in estimates are (1) to gain more usable and realistic records of class characteristics and (2) to develop improved techniques for applying the consequent knowledge to predictions in appropriate confidence settings. The current state-of-the-art in reliability predictions rests at the developmental level of these data and techniques, and much room remains for advancement.

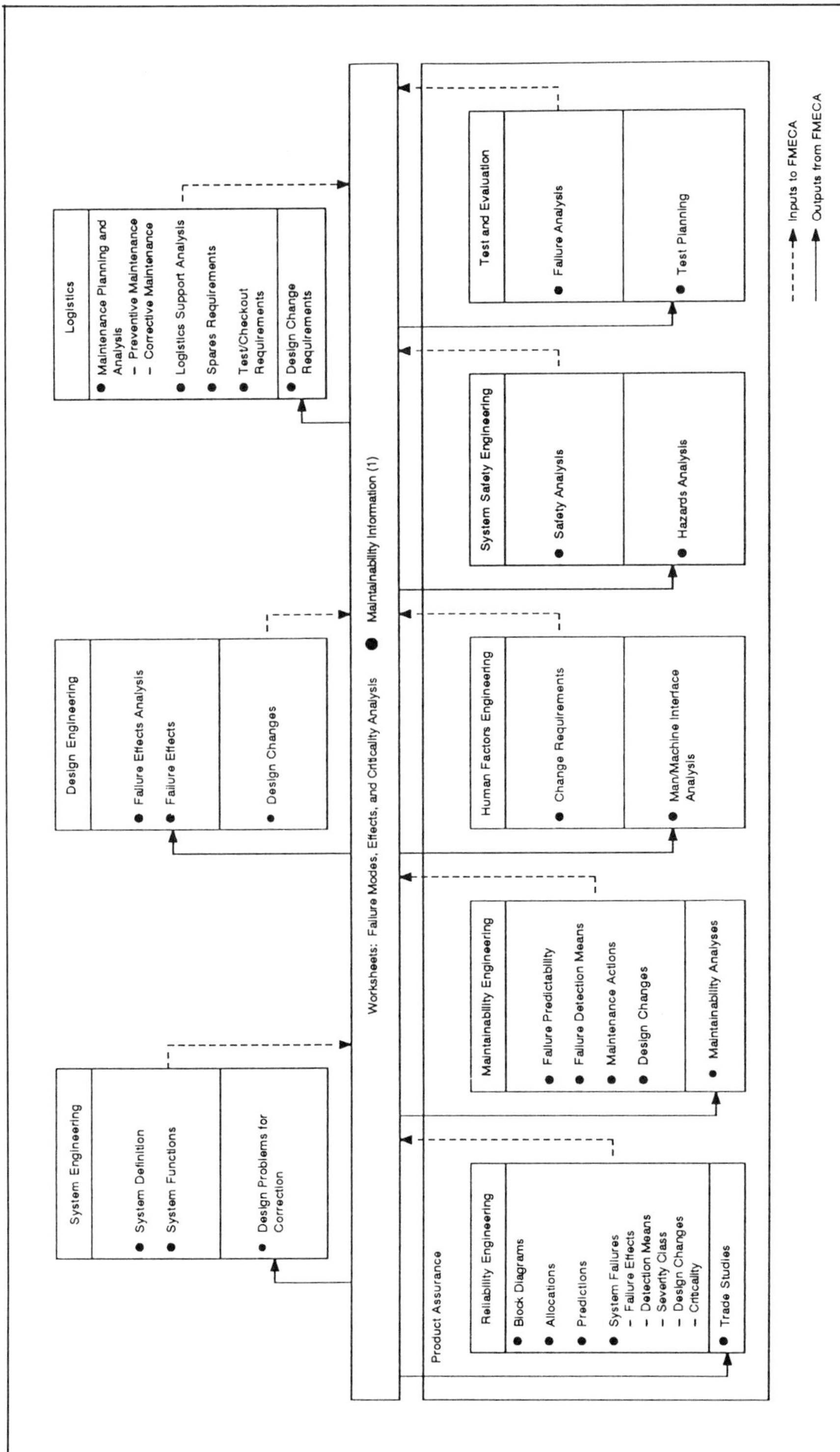

Fig. 5.6 Failure modes and critical effects flow. [1]*IAW MIL-STD-2080A Paragraph 5.2 and Figure 5.*

Fundamental limitations of reliability estimates

The art of estimating the reliability of electronic products has fundamental limitations. It would be too much to ask that reliability be predicted exactly. In fact, it has been shown that this would be philosophically undesirable since an absolutely exact prediction of the future is an immutable future and, as was pointed out, it is the business of electronic design and reliability engineering to design the future, not to take it as it comes.

Beyond such philosophical considerations, however, are practical limitations (such as those that depend on data gathering and technique complexity). Considerable effort is required to generate sufficient data to report a statistically valid reliability failure rate for a part. Casual data gathering on a part class occasionally accumulates data more slowly than the advance of technology in that class; consequently, a valid level of data is never attained. In the case of many part classes, the number of people participating in data gathering all over the industry is rather large, with consequent varying methods and conditions which prevent exact coordination and correlation. Also, part reliability in the field use of equipment is difficult to examine due to the lack of suitable data being acquired. Thus, it can be seen that deriving failure rates (mean values) is empirically difficult, and obtaining valid confidence values is practically precluded because of the lack of correlation.

Failure rate data obtained from field use of past products is applicable to future concepts, depending on the degree of similarity in the hardware design and anticipated environments. Data obtained on a product used in one environment is not necessarily the same as data that would be obtained in another environment. The stress imposed by a new environment may alter the data, especially if the new stress exceeds the design capabilities. Other variants that can affect the stated failure rate of a given product are different uses, operators, maintenance practices, measurement techniques and definitions of failure. When considering a comparison between similar but unlike products, the possible variations are obviously even greater.

Thus, a fundamental limitation on reliability estimates is the ability to accumulate data of known validity for the new application. Another fundamental limitation is the complexity of estimation techniques. Very simple techniques omit a great deal of distinguishing detail, and the estimate suffers from inaccuracy. More detailed techniques can become so bogged down that estimates become costly and may actually lag behind the principal hardware development effort.

Failure analysis

Failure analysis is a very important tool used by the reliability engineer to determine the cause of failure during the advanced development through production phases of a contract. Fig. 5.7 indicates the flow in which failure

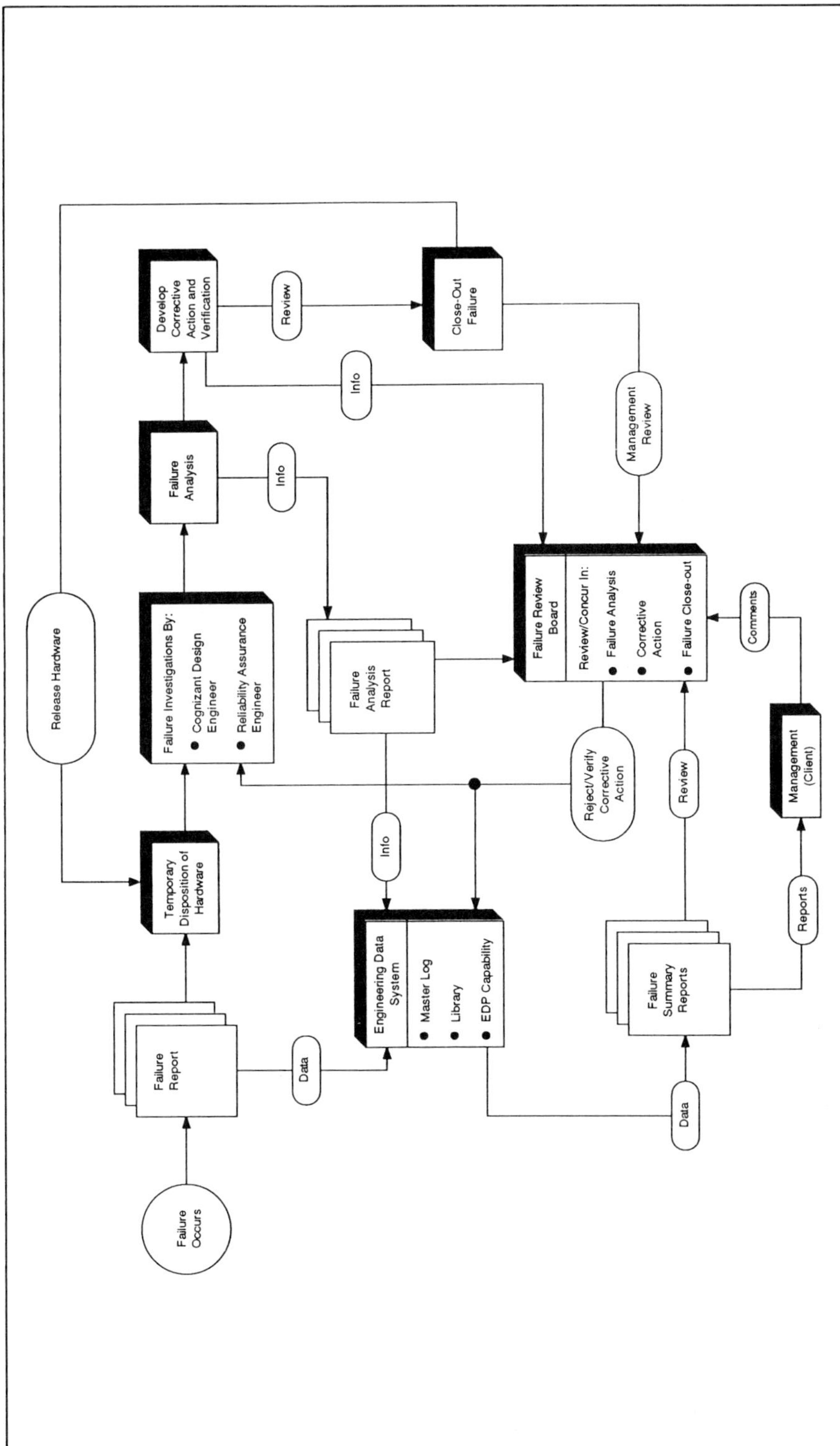

Fig. 5.7 Failure analysis. EDP, electronic data processing.

reports travel among all functions in the organisation. The failure report is the basis by which the reliability engineer performs all investigations. It is done in conjunction with all of the other functions within the organisation. The idea is to be able to establish a data base that identifies the problems occurring from the initial portion of the design through its production and deployment phases. The reliability engineer uses this tool to indicate to the design engineer that there may be problems with the design. The paperwork associated with the failure report and the failed item are turned over to the reliability engineer or the quality engineer to determine the actual cause of failure. This is now being called 'forensic engineering'. The part is analysed, and the cause of failure is determined. If a sufficient number of similar failures occur, the reliability engineer has a basis for corrective action. He will prepare a corrective action report, notify the engineering staff and programme management office, and indicate to them that there is a possible latent defect in the design.

This then becames a source for further investigation, and a determination of the type of action that is required to eliminate the problem. It will then be adjudicated between the engineering staff and the reliability group. Failure analysis is the mechanism by which the reliability engineer closes the loop to ensure that all of the failures that have occurred have been resolved. If resolution cannot be accomplished, it goes into an active file, and failure trend analysis is then started to determine whether similar failures occur. If so, the reliability engineer would go back to studying the failure modes and critical effects analysis to determine in which part of the functional aspects of the design this device resides, and to decide whether there could be an actual flaw in the design. If a flaw is found, it either is sent back to the manufacturer for complete retrofit, or to the engineer to determine how best to work around the problem. Either way, there is a complete resolution.

To have a full appreciation of what the reliability programme is and how it operates, Fig. 5.8 demonstrates the aspects of all of the material generated by the reliability engineer, and how it moves between and among the various functions within the organisation. It starts off with the reliability programme and then follows the actions necessary to support the tasks being performed by the reliability engineer. For the sake of brevity, a number of tasks inside each of the boxes have been combined, to show that certain elements must be performed. These elements, or tasks, are essential to the reliability engineer in support of the programme effort. The flow is generic, but has a direct impact on how the various elements relate to one another.

Maintainability

The trend over the past years has been to build increasingly complex systems with quicker reaction time and improved accuracy. This complexity requires more equipment, which results in a lower MTBF. With in-

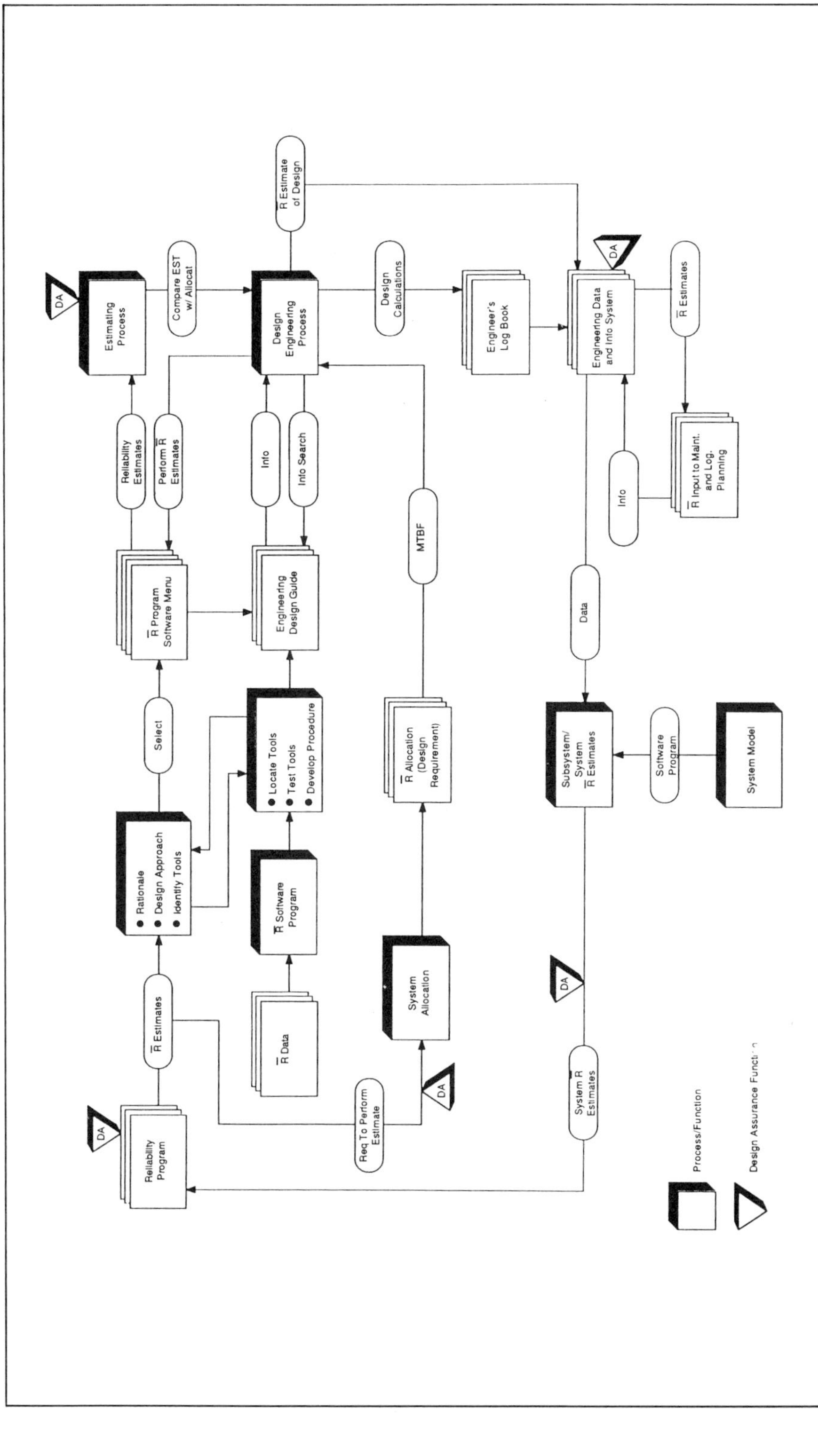

Fig. 5.8 Reliability programme flow.

creased hardware complexity and lower MTBF, satisfactory maintenance becomes a greater problem. A derivative of this could be more maintenance personnel to perform the additional tasks and more facilities than available space will accommodate. Increasingly complex maintenance tasks could require skilled personnel who may not be readily available. Various approaches must be taken to help to reduce this problem. They should be discussed during the onset of any programme. Proposed solutions to a maintainability problem should be evaluated in terms of the total product.

Design engineering and analysis techniques should be used to identify the potential maintainability problems and to make the fundamental trade-off decisions necessary to achieve the optimum balance among the design requirements, logistics plans and costs. An arbitrary decision could easily result in a more serious and costly problem. Thus, design orientation is needed to solve the simplest maintainability problem.

Before describing maintainability in its total environment and its relationships to reliability and logistics, a definition of maintainability is required. The following definition is used for discussion purposes: maintainability is the ease and rapidity with which a defective item can be restored to its operation condition. It is a function of product design installation, availability of personnel, adequate maintenance and test equipment and its environment. Therefore, a user's measure of a product's basic worth is its effectiveness, logistics supportability and maintenance costs. All three factors are largely dependent on the product availability and maintainability. Maintainability is therefore a critically important consideration when planning for a new system.

Programme applications

While maintainability programmes have been required for most of the military efforts over the last few years, they have not received the same attention as reliability programmes. When maintainability has been applied, it has not resulted in the objectives it was designed to attain. Factors contributing to this condition include voids in the technology (which have not been eliminated in the past thirty years) and misapplications of technology (primarily in integrating it into the design process). With the increase in the complexity of modern products and their related economics, these factors can no longer be ignored. Another major concern has been the management support and motivation associated with maintainability. Management, including programme and technical managers, currently do not have a sufficient understanding of the differences between maintainability and maintenance.

This misunderstanding affects the application of maintainability to programme objectives, resulting in high maintenance and life cycle costs. Maintainability is a design discipline that must be integrated through the design engineering process. If the objectives of the maintainability programme are to be realised as the product complexity increases, maintenance costs will increase exponentially, as will availability and life cycle costs. Management must begin to recognise the technical attributes of the methodologies and economics involved in achieving technical and management objectives on major programmes. Conversely, technical personnel must become aware of and develop techniques and technologies that will economically achieve the programme's or project's technical objectives.

As with reliability, maintainability activities must be described early so that the maintainability engineer can progressively increase his involvement as the design grows. For argument's sake, let us describe and discuss the specific aspects associated with maintainability.

The elements that should be considered are:

- Maintainability analysis.
- Maintainability predictions.
- Logistics support analysis (LSA).
- Design reviews.

Each of these elements will be further described in this chapter. In an attempt to ensure that the information is properly sequenced, a series of events necessary to support the activities have been structured. First and foremost, maintainability's role in design, like reliability's role, must be immediately recognised. But maintainability does not become involved with the design until later in the product life cycle. However, maintainability plays a critical factor in determining how and where modules will be located in the configuration of a product.

Prior to any detailed definitions of maintainability, a brief discussion of maintenance is appropriate. (More discussions on maintenance and the maintenance concept will be found in Chapter 7). The maintenance concept is developed early in the design phase. It is essential that a maintenance philosophy or concept be imposed so that, as the product designer begins to structure his functional design, sufficient room is allocated for the development of diagnostics. From the product's operational requirements, the quantitative maintainability requirements can be derived. These maintainability requirements provide the necessary information to establish the logistics support planning criteria. The maintenance concept, which defines the scope and proposed methods of repair at each level of maintenance, attempts to satisfy the quantitative maintainability requirements of logistics support.

The primary objectives in developing appropriate maintenance are to provide realistic, definitive and uniform product design in support planning. The rapidity with which the maintenance technician accomplishes his job will be directly applicable to the amount and degree of diagnostics included in the product. Depending upon the size and complexity of the product in question, the diagnostics will routinely check the major units. It will also depend upon the amount of room available in the computer. Utilising the appropriate diagnostic routines, the design engineer and the maintainability engineer will be able to allow the computer process to diagnostically troubleshoot the equipment down to the lowest possible level. Application of this technique will assist in simplifying the amount of work that the maintenance technician must perform. Diagnostics may be run either online or offline, depending upon the availability and need of the product in question. Maintenance, therefore, should be considered one of the three sides of the triangle which ultimately supports the requirements associated with ILS.

Once a software diagnostic package to support the maintenance concept has been defined, further refinement will be applied when the reliability engineer completes his reliability studies. With the identification of critical parts, the routines by which the computer program will be run will be directed toward the higher probability of failure. It then becomes the maintainability engineer's responsibility to ensure that these components are easily accessible in the design. It must be recognised that reliability becomes actively involved in the design earlier in the design phase than maintainability.

The reason behind this is that during the functional layout of the components and the partitioning of the equipment, reliability must be measured immediately so that the impact on the design can be quickly felt. The reliability engineer, in general terms, gets involved during the development and validation phase of a design. This sets the guidelines for the full-scale engineering development (FSED) phase. Reliability affects the functional operations of design, where maintainability affects the form and fit of the design rather than the function. All three – form, fit, and function – come into their own during the FSED phase of a programme. Maintainability, therefore, plays a more important role later in the design phase.

As with reliability, maintainability is directly affected by the efforts being performed by the design engineer. Therefore, Table 5.3 has been constructed that identifies maintainability's role in design. The identification of the tasks to be performed, the organisational responsibility for those tasks, and the specific maintainability engineering roles (vis-à-vis providing inputs, reviewing documentation, supporting specific developments and co-ordinating results of their products) must be recognised. If these are done in consonance with what the reliability engineer does during the FSED phase, the job of the maintainability engineer becomes a little easier.

The maintainability engineer's job is no less important than any other function involved in the design of a product. Therefore, maintainability should be compatible with the maintenance concept in the support plan. Conversely, the support plan and the maintenance concept for the product should be adjusted during the requirements or early analysis to be compatible with the existing design constraints.

The interfaces between maintainability engineering and design and between maintenance engineering and support must be considered early in the programme. Coordination of reliability and maintainability requirements in a summary of the maintenance concept should be part of the early development phase. This coordination is required to facilitate visualisation of the interface, mutual interactions and potential trade-offs.

Fig. 5.9 shows the relative importance of programme functions and events in achieving maintainability requirements. This chart depicts the various elements that must be performed during the early phases of a specific programme. Most of the elements shown here are required to ensure that maintainability is an achievable entity in itself. This is also depicted in Table 5.3, which reflects the necessary roles of maintainability in design. It is essential to recognise that not all of these elements must be incorporated. They should be tailored. The degree of tailoring will be commensurate with the amount of work that must be performed on the design. The aspects of tailoring are not necessarily self-defining, but the maintainability engineer must have an appreciation of the depth and breadth to which he must exercise his engineering prerogatives.

To give the logistics manager and the logistics elements managers a better appreciation of what it takes to perform the maintainability design role, we have decided to definitise each of the four major elements described on page 73. This does not preclude the logistician or logistics elements managers from participating in the activities associated with maintainability engineering, but it gives them the necessary insight required to support the activities being performed and to assist the maintainability engineer in

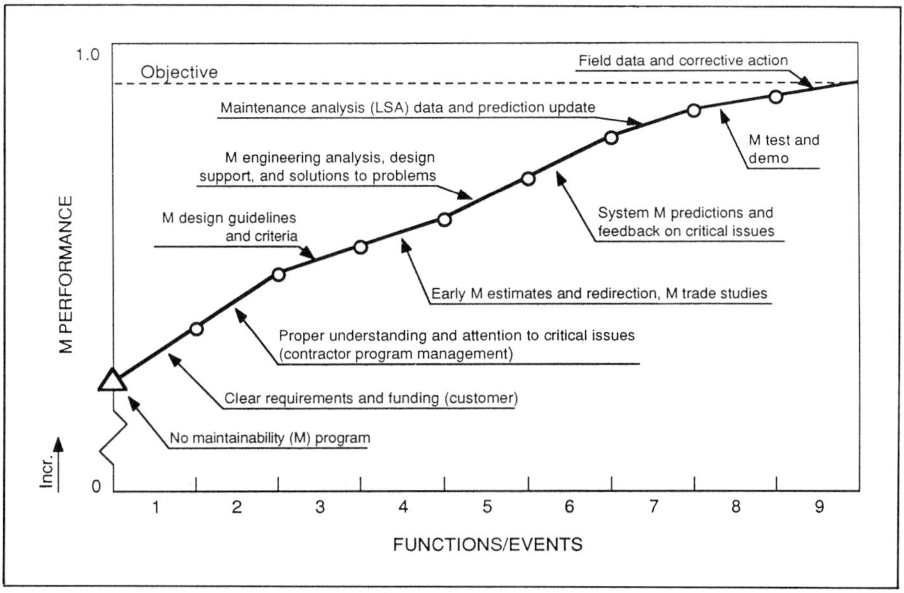

Fig. 5.9 Relative importance of programme functions and events in maintainability achievement.

Table 5.3 The role of maintainability in design

Task	*Responsible organisation*	*Maintainability engineering role*			
		Provide	*Review*	*Support*	*Coordinate*
Hardware design	Engineering	• $\overline{M}$ design guidelines • $\overline{M}$ requirements/allocations • Bit/Bite requirements • $\overline{M}$ analysis results • Math models • $\overline{M}$ prediction per MS 472	• Designs for – Repair access – Failure impact – Compliance with $\overline{M}$ requirements	• LCC analysis • Trade-off studies • $\overline{M}$ risk assessment	• $\overline{M}$ model use – Proper $\overline{M}$ cost choices based on LCC
Reliability design	Engineering	• Bit/Bite false alarm potential • FMECA-MI	• FMECA • Failure rates (system and component) for $\overline{M}$ input	• Trade-offs to select optimum $\overline{M}$ combinations • Bit/Bite applications t/o decisions	• $\overline{R}$&$\overline{M}$ quantitative and qualitative requirements
Software design	Engineering	• $\overline{M}$ design guidelines • Standards to maximise M of software	• Logic for failure modes/effects	• Software design requirements	• Modelling of software faults
Quality assurance provisions	Through ILS manager	• $\overline{M}$ test requirements • $\overline{M}$ failure reporting requirements	• Failure data • Test results from incoming inspection, parts screening, in-process screens, ATPs	• Procedure development	• QA processes

Table 5.3 (continued)

Task	*Responsible organisation*	*Maintainability engineering role*			
		Provide	*Review*	*Support*	*Coordinate*
Supplier management	Procurement	• $\overline{M}$ programme requirements • Quantitative $\overline{M}$ requirements for assigned HW/SW	• Supplier – Failure data – $\overline{M}$ planning – $\overline{M}$ programmes	• Contract negotiations on $\overline{M}$ terms/conditions	• Supplier implementation of MHC $\overline{M}$ programme
Safety design/ Analysis	ILS	• FMECA – MI • $\overline{M}$ design guidelines	• Hazard analysis for impact on $\overline{M}$ design	• Trade-off studies	• Human factors and maintenance design
Human factors design	ILS	• $\overline{M}$ design guidelines	• Impact of human factors design	• Trade-off studies	• Man/machine interface requirements to ensure adequate failure prevention/ detection/correction
Logistics engineering	ILS	• $\overline{M}$ design guidelines • FMECA – MI • $\overline{M}$ data (for LSA)	• Maintenance planning for potential failure inducing actions • Packaging concepts	• Maintenance planning • LSA • Analysis of packaging impacts on $\overline{M}$	• LSA • TAT
System test	ILS	• $\overline{M}$ test requirements • $\overline{M}$ test plans • $\overline{M}$ test criteria • Test conduct	• Test procedures • Test results	• Failure analysis	• $\overline{M}$ Demo

successfully accomplishing his role. Logisticians must recognise that maintainability has a direct impact on what occurs within the logistics environment. If the maintenance concept is not properly identified, and the locations of assemblies, units or equipments are not properly placed, then the role of the logistician becomes much more difficult. He is not acting as an auditor in the true sense of the word but as a policeman in attempting to stop work that will have an adverse effect on everything that preceeds the early portions of the design.

Maintainability analysis

As in reliability, maintainability engineering should perform an analysis to determine what the various allocations of maintainability will be among all the equipment items, units, assemblies or modules that will be incorporated into the design. The maintainability engineer performs this so that he has a basis from which he can determine the relevant impacts on each of the units as the design progresses. The logistics engineer, with the maintainability engineer, should ensure that the diagnostics proposed early in the design are applicable and serve as the tool by which the entire product can be checked. The depth, breadth and degree of diagnostics will be determined based on the funding available, the speed at which the maintainability requirements have been identified and the level of complexity at which the maintainability engineer has to work.

To properly analyse the maintainability aspects of a programme, logisticians should understand the actual flow of data between and among the various elements within the functional components of a corporation. Maintainability *per se* follows generally the same flow as reliability. The movement of data between the logistics engineer, the design engineer and the maintainability engineer is depicted in Fig. 5.10, which identifies the flows among the relative functional activities needed to ensure that maintainability is appropriately reviewed and evaluated during the development of a product's life cycle. The maintainability analysis flow is not necessarily time oriented, but it reflects how best to address maintainability and what the interfaces are during the period of involvement.

Maintainability predictions

The maintainability prediction is the tool by which the maintainability engineer determines how long it takes to localise, isolated, dissassemble, reassemble, align and test a product after a failure has been identified. The isolation and localisation are predominantly done via the diagnostic routines. In more complex techniques, the diagnostics are run so that they aid

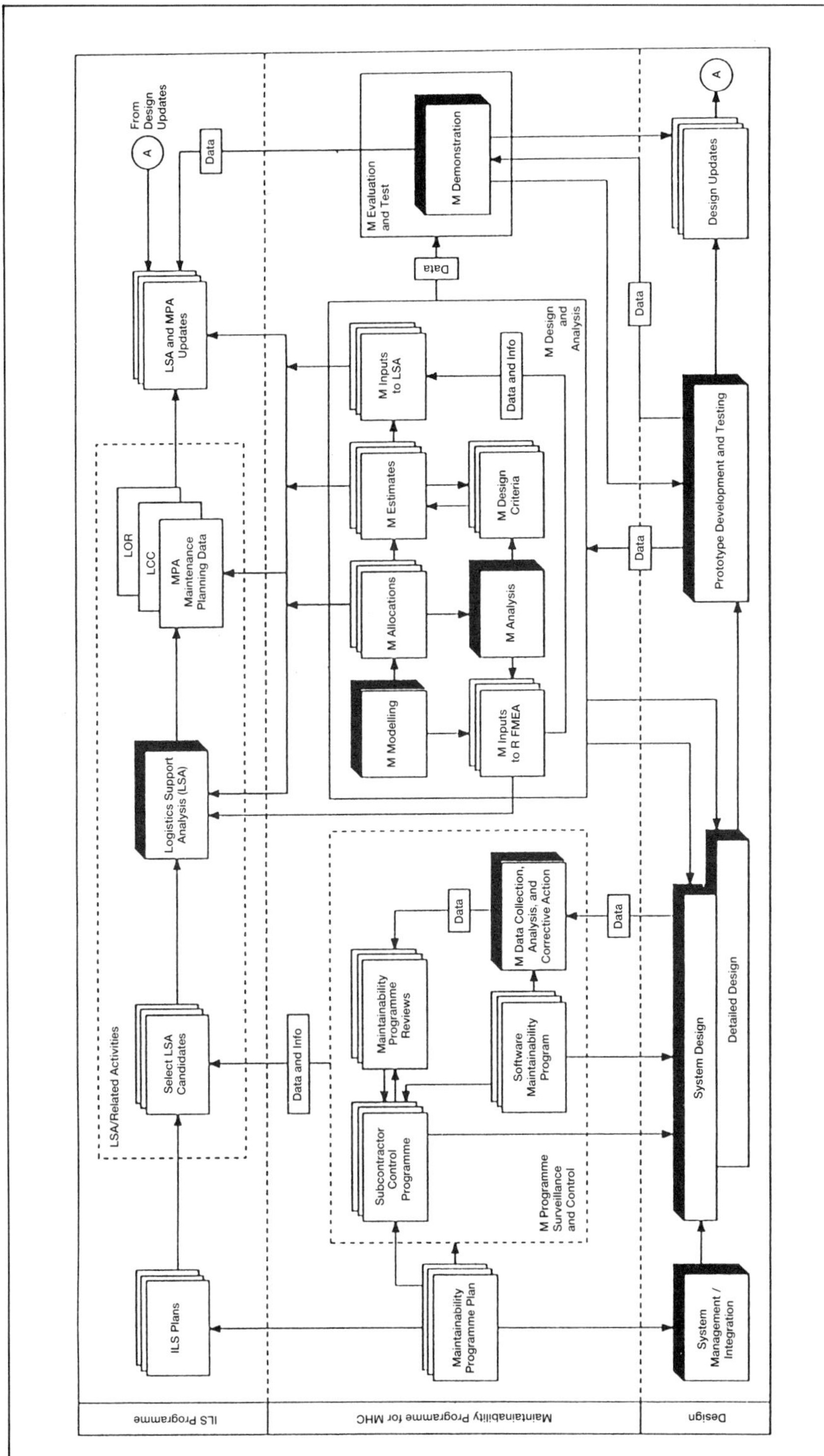

Fig. 5.10 Maintainability analysis. FMEA, failure modes effects analysis; LCC, life cycle cost; LOR, level of repair; M, maintainability; MPA, maintenance planning analysis.

in the simplification of troubleshooting performed by the maintenance personnel. The dissassembly, reassembly, alignment and test are performed with human involvement. The test is a reaffirmation that the failed part has been properly replaced, and the diagnostics routines are used to verify that the product is back in operation. This tool, or predictive technique, is used to ensure that the allocation of maintainability requirements has been achieved.

If the values that are calculated are in excess of the established allocations, the maintainability engineer is required to review the process by which the troubleshooting techniques are applied and to determine (1) if the diagnostics routines are successfully doing the job, (2) whether the location of the failed part is appropriate for removal or replacement and (3) whether the appropriate test equipment is available when needed to help troubleshooters ensure that the part being removed is indeed the failed part. The maintainability prediction use the results of the reliability prediction because this is the basis by which the frequency of failure is measured. The maintainability prediction is performed early in the FSED phase, and is normally presented as part of one of the major design reviews. If problems associated with the maintainability predictive techniques identify that maintainability is not being achieved, the results should be re-analysed and corrective action should be initiated to ensure that maintainability becomes an achievable requirement. The results are proven during a maintainability demonstration which is performed at the conclusion of FSED. The maintainability demonstrations, as well as reliability testing, will be discussed further in Chapter 8.

Maintainability – its role in logistics support analysis

Maintainability and reliability have direct inputs into the LSA. They become the driving forces for all of the following data sheets represented in MIL-STD-1388-2. They serve as a basis from which the LSA process will calculate such items as provisioning, maintenance man hours, allocation of maintenance tasks and specific support equipment, and are input into the B-1 and B-2 sheets, representing reliability and maintainability. This is the first set of sheets that are established as part of the LSA process. It is important to know early that the procuring activity should prepare the 'A' sheet and supply it to the logistician. The 'A' sheet essentially represents the mission times, reliability requirements and other pertinent data needed to support the establishment of appropriate reliability and maintainability data for the programme. If this sheet is not supplied as part of the contractual package, the logistician and the project manager should request its generation by the procuring activity. Without the sheet, it becomes almost impossible to satisfactorily perform the remainder of an LSA task.

Therefore, when evaluating a request for proposal, the logistician should immediately collect the necessary information and obtain the 'A' sheet so that any predictive methods or associated techniques are readily available to him. Without the sheet, it will be difficult to achieve a balance between readiness, operational capability and cost. To further define this, Fig. 5.11 (a) and (b) have been included to help illustrate what the maintainability analysis in its appropriate flow is like. This maintainability analysis is directly related to Task 205 of the maintainability standard (MIL-STD-470). A paper by William R. Downs (Downs Technical/Management Services), describes the specific flows of Fig. 5.11 (a) and (b). The following excerpt is from his Report on Basic Maintainability.

> The author considers this test the most important of the twelve specific tasks in achieving maintainability performance objectives in equipment. It may be the least understood programme element within the industry. [Fig. 5.11 (a)] shows the maintainability analysis in relation to the LSA in order to delineate the two in regard to the interfaces and their differences. The two separate programmes are sometimes confused.
>
> It is considered that maintainability analysis, represented by the upper group of tasks, is the optimising of the maintenance capability in the design of the operational equipment required to perform the primary mission for the system, while the LSA (the lower group) is concerned with optimising the supportability of the same equipment.
>
> As one follows the maintainablity analysis flow, it includes trade studies and support and functional analysis of the systems, leading to the repair policy decisions, which are evaluated in relation to the similarly developed support policy to arrive at a compatible maintenance concept. The optimum repair policy is essential to the design evaluations, the development of design criteria, and the allocation and prediction models. The design is evaluated in trade-offs considered to improve design or verify that design will meet the criteria. This evaluation is documented as the maintenance repair method analysis performed in close support of the technical design. Live replaceable units (LRUs) are identified as online repair policy. The maintenance tasks and time data are developed from these analyses. These are used in the maintainability predictions. Design improvement recommendations are fed back as appropriate, based upon the qualitative and quantitative results. The maintainability analysis data are provided to the organisation performing LSA for their information in performing the maintenance and support analysis, which may provide its own design improvement recommendations.
>
> The LSA supportability analysis is being performed within the same time frame. This includes level of repair and predictions of costs to support personnel and training. It is considered that the analysis shown in this figure is complete in time for the preliminary design review (PDR), with the maintainability engineer reporting on the predicated down times for the specific levels of maintenance repair methods required to achieve the desired maintainability performance, and problems and methods for resolution if applicable. The LSA engineer will be making similar reports regarding supportability concerns.

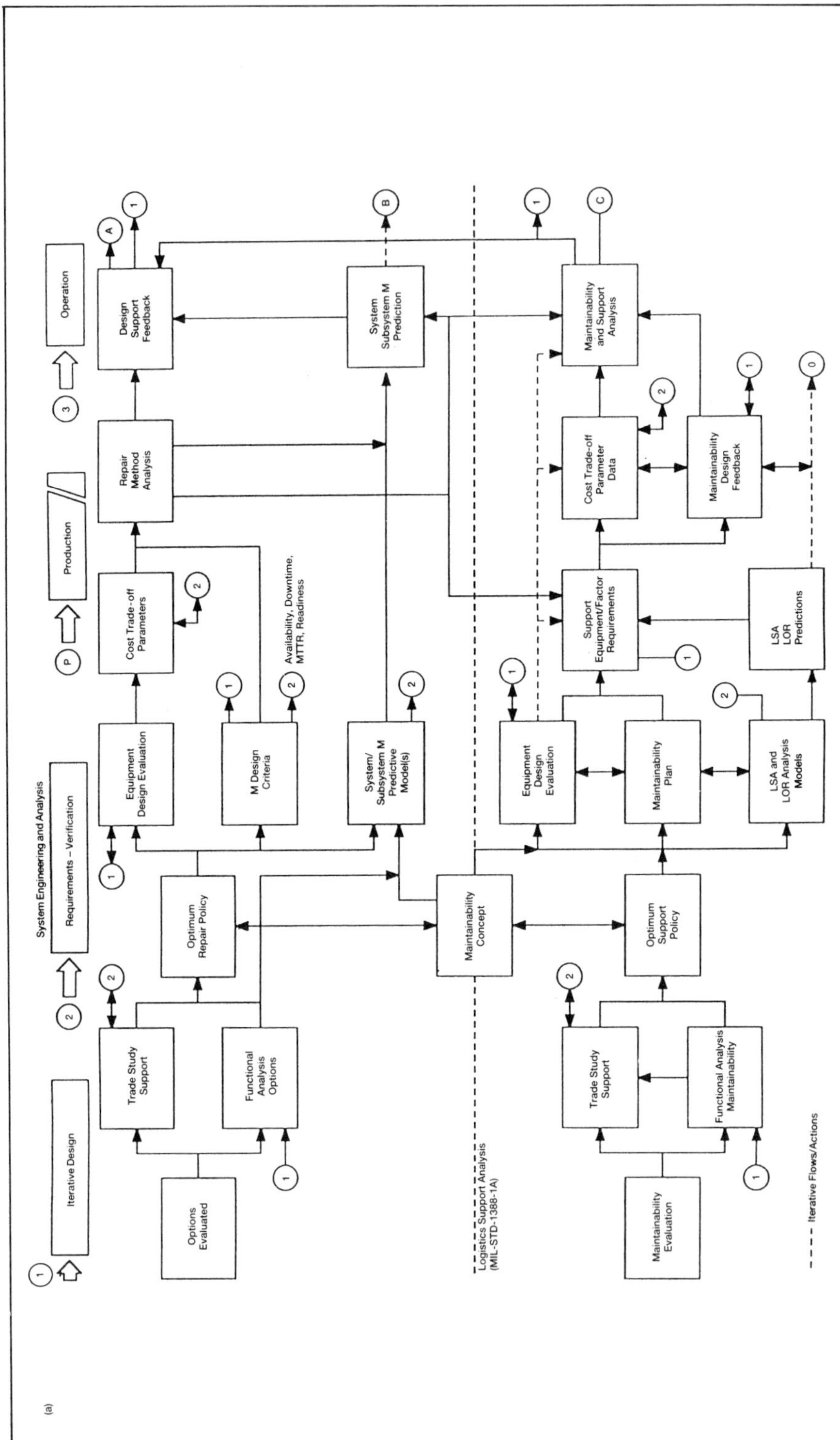

Fig. 5.11(a) and (b) Maintainability analysis/logistics support analysis (LSA) interfaces. LOR, level of repair; MTTR, mean time to repair.

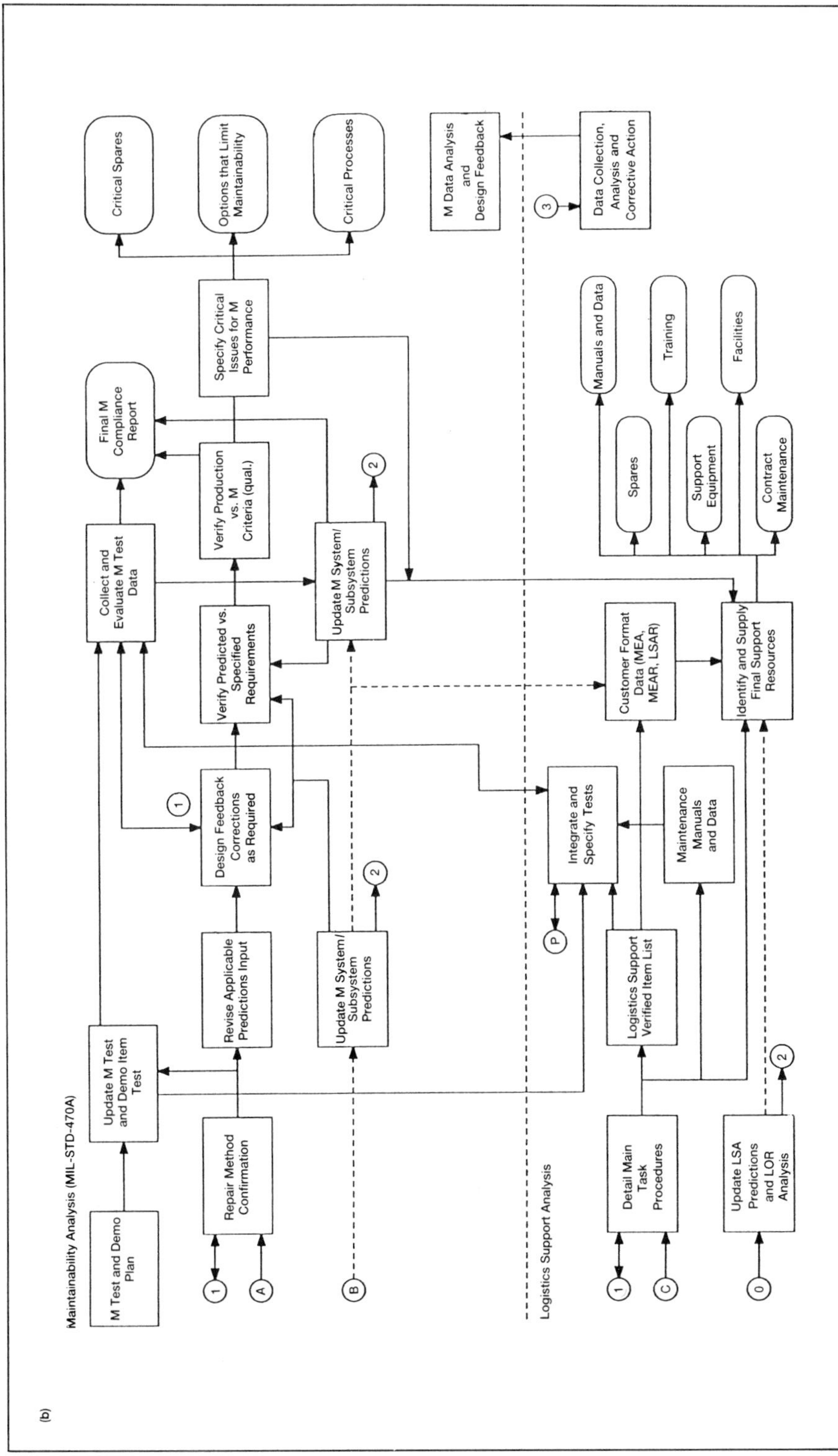

Fig. 5.11 — Contd.

> [Fig. 5.11 (b)] shows the continuation of these analyses after the PDR time frame until hardware delivery. At this time, it is considered that the maintenance tasks and time data of the LSA programme, based upon the released designs, will be used by maintainability engineering as an improved evaluation of the current design status. These data will be reviewed to confirm that the repair methods are still valid, and will be input to the predictions to replace the earlier maintainability time elements. Note that this is the second time evaluation, and occurs during the post PDR time period. If this were the only analysis performed in this time frame, it could not be relied upon to provide the level of design improvement required by the maintainability programme.
>
> The maintainability test and demonstration plan may require updating. In any case, demonstration items will be specified. [This information will be covered as part of the test portion of this book (see Chapter 8).]
>
> The maintainability outputs shown at the end of this flow include the final reports and recommendations that may be appropriate regarding any issues critical to achieving their required maintainability performance. This might include recommendations for sparing levels or spare item configurations to minimise replacement times. It may be important to recommend sequences of operations that should be employed to minimise the constraints on personnel performing time-critical maintenance tasks, critical processes that may be controlled because of their influence on down time, and so on. These conditions must be made known to the personnel responsible for the manuals and data, support equipment, training, etc., that evolves out of the LSA data process.[4]

Some US Government military documentation indicates that maintainability analysis is based on the FMECA discussed earlier in this chapter. Actually, the maintainability analysis is a comparable activity which is performed in the same time frame.

Maintainability predictions

To ensure that established requirements for maintainability have been met, it is necessary to assess the product's maintainability characteristics periodically throughout the development process. Two principal techniques are available for this process: predictions and design reviews.

During the planning phases, the predicted maintainability of different products proposed to meet an operational need is a critical factor in selecting the optimum course of action. Since a limited quantity of specific data is available in this early phase, maintainability predictions are largely based upon experience with predecessor equipments and on prediction techniques applicable during the planning phase. As mentioned earlier, maintainability predictions are rarely developed other than by establishing the

---

[4] Downs, William R., *Report on Basic Maintainability* (Downs Technical/Management Services).

maintainability allocation that was discussed in this chapter. During the design phase, reliability and maintainability predictions can be employed to determine the inherent availability of the proposed product, the effects of proposed changes on availability (if availability is applied to the design) and the optimum trade-off of equipment characteristics. Predictions made during this phase are generally more accurate than those made in the planning phase, since more specific information is available.

Because they are more accurate than assessments conducted in the beginning phases of a design, the predictions obtained should be used to upgrade earlier predictions and the techniques used to obtain them. The prediction techniques were developed principally for electronic equipment. The rapid changes in electronic technology necessitated the constant updating of these techniques. In addition, prediction procedures – particularly those applicable during the early phases – are not as accurate as the analysis procedures available to other aspects of the engineering process. However, since they are based upon empirical data, they provide useful estimates of maintainability characteristics of a piece of equipment during early design phases. MIL-STD-472 describes approaches to maintainability predictions during the equipment life cycle.

Basic assumptions and interpretations. Each maintainability prediction procedure included in the military handbook depends upon the use of recorded reliability and maintainability data and experience which have been obtained from comparable systems and components under similar conditions of use and operation. It is also customary to assume the applicability of the 'principle of transferability'. This assumes that data which accumulate from one system can be used to predict the maintainability of a comparable system which is undergoing design, development or study. This procedure is justifiable when the required degree of commonality between systems can be established. Usually during the early design phase of the life cycle, commonality can only be inferred on a broad basis. However, as the design becomes refined, during later phases of the life cycle, commonality is extendable if a high positive correlation is established relating to equipment functions, to maintenance task times and to levels of maintenance. Although the techniques contained in this handbook have been proposed and appear to fit certain applications, it should be borne in mind that they have not truly been tested for generality, for consistency one to another or for most other criteria dealing with broad applicability. It should also be borne in mind, though, that experience has shown that the advantages greatly outweigh the burden of making a prediction. For that reason, it is not the purpose of this document to deter further research or inquiry.

Elements of maintainability prediction techniques. Each maintainability prediction technique utilises procedures which are specifically designed to satisfy its method of application. However, all maintainability prediction methods are dependent upon at least two basic parameters:

- Failure rates of components at the specific assembly level of interest.
- Repair time required at the maintenance level involved.

There are many sources which record the failure rate of parts as a function of use and environment. The major advantage of using the failure rate in maintainability prediction calculations is that it provides an estimate of the relative frequency of failure of those components which are utilised in the design. Similarly, the relative frequency of failure of components at other maintainable levels can be determined by employing standard reliability prediction techniques using parts failure rates. Failure rates can also be utilised in applicable regression equations for calculating the maintenance action time. Another use of the failure rates is to weight the repair times for various categories of repair activity, in order to provide an estimate of its contribution, to the total maintenance time.[5]

Maintainability programme flow

Fig. 5.12 will assist the logistician in understanding the applications of maintainability. It illustrates the flow of technical data and information among the relevant maintainability tasks to be performed by the maintainability engineer. It ties in the total maintainability programme from the initiation phase of actually writing a programme plan, to the conclusion at the point at which a maintainability demonstration has been performed. The various steps within each of the specific facets of maintainability have been identified.

Design reviews as they apply to reliability and maintainability

The design review meeting scheduled for any design programme should include at least a preliminary design review, detailed analysis review and final design review. If testing is required, we should consider a Test Correlation Review. We will begin by defining what we believe to be absolutely necessary to support each of these major reviews.

1. *Preliminary design review (PDR).* Several alternative designs for meeting requirements are generally available. The primary purpose of the PDR is to make a choice from among alternative design approaches that may have evolved during the design process. The choice should be one of the following, in order of preference:

- The simplest design that meets both reliability and maintainability requirements.
- The design that has the highest maintainability and the most accurate reliability.
- The design that shows the greatest promise of meeting both the reliability and maintainability requirements.

---

[5] US Department of Defense (1984) *Maintainability Prediction* MIL-STD-472.

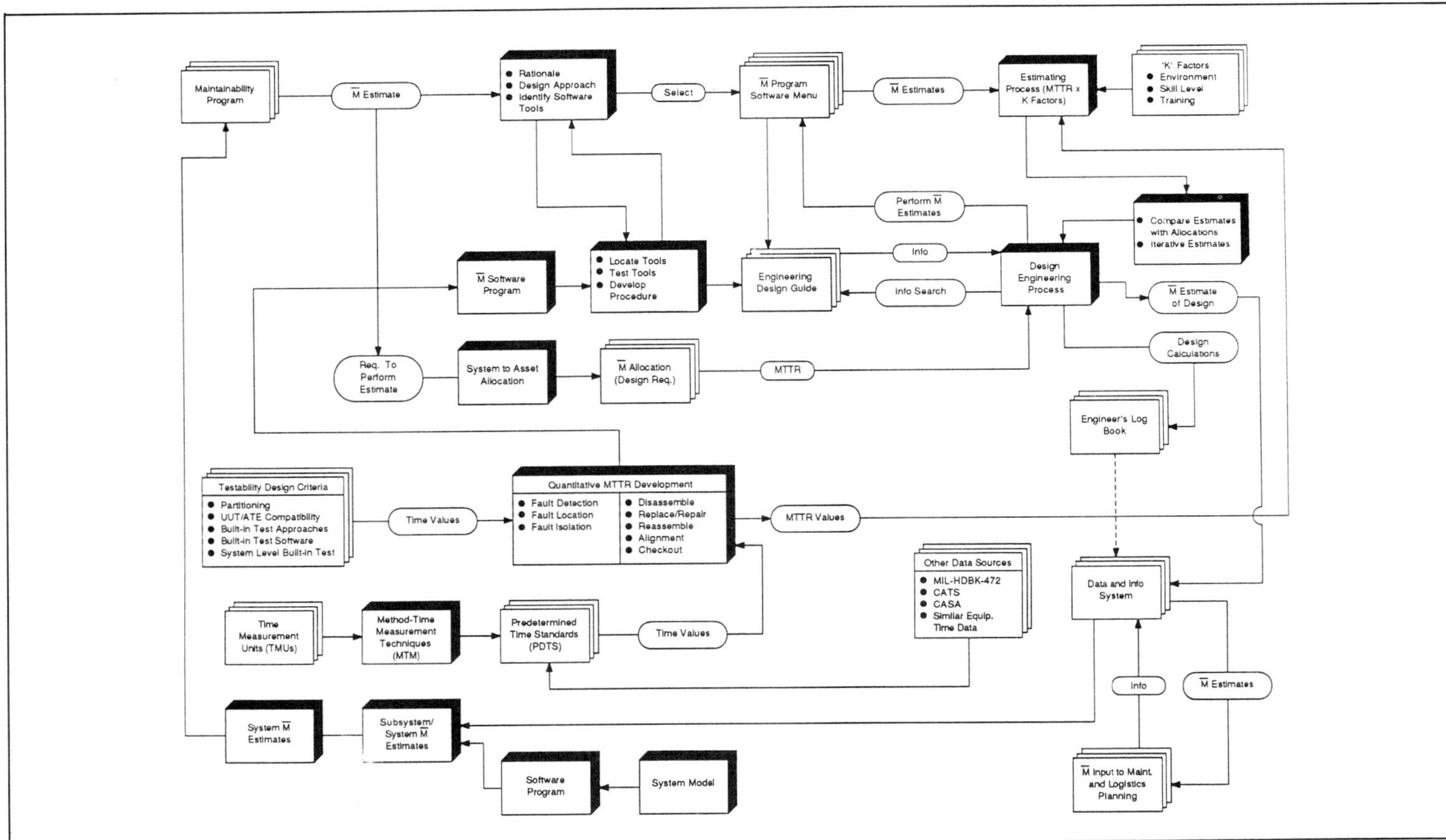

Fig. 5.12 Maintainability programme flow. MTTR, mean time to repair; UUT/ATE, unit under test/automatic test equipment.

The results of this first design review should include an understanding of the weak areas in the chosen design. The reliability and maintainability block diagram of the chosen design will show the series and parallel efforts that should also result from the review. The PDR should also reveal any lack of data or need for more design data, such as:

- preventive and scheduled maintenance requirements;
- more information on hardware construction and accessibility;
- diagnostics and testing schemes;
- special facilities that may be required;
- techniques to ensure that reliability achieves the specified requirements;
- functional differences between what was originally designed and the initial design review;
- changes in the equipment environment that will affect reliability analysis;
- functional design changes that will affect the probability of achieving the reliability goals.

2. *Detailed analysis review.* By the time the design has progressed through the initial stages, parts have been estimated and selected and part applications have been determined. At this point, with only one design to consider, a thorough reliability and maintainability analysis can be performed. To determine whether the design will meet both the reliability and maintainability requirements, a maintainability prediction and a reliability estimate are performed.

If the predictions and estimates indicate that reliability and/or maintainability will not be met, then management must decide whether to abandon the present design and begin again, or to concentrate on ways to improve the design, or to establish techniques by which parallel operations can be performed without having a major impact on the design. These decisions are made during a detailed analysis review. If improvement is needed, areas requiring more attention should be identified. This is the point at which the following design decisions are required: (1) whether redundancy is required versus rapid fault isolation techniques, or (2) redesign of inaccessible areas is required versus a source of highly reliable parts. The latter is a typical example of an extensive interface between reliability and maintainability.

Planning should precede the meeting to ensure that the design review is tailored to the design. Any misapplications should be identified in the meeting. Questionable areas, such as those in which severe environmental conditions appear to be troublesome, should become evident. These problems may be identified and earmarked for subsequent attention under the category of 'design for reliability'. In analysing the results of this design review, management should determine whether the decisions made in the previous design review are valid and how to plan in the continuation of the design phase. If invalid decisions have been made in earlier design reviews and the design has regressed to the extent of having a major impact on the ultimate ability to accomplish the mission, it is essential that the management team quickly assesses the impact of the change. Then they must determine the best approaches to ensure that the achievements of the required specifications can be met with redesign.

3. *Final (critical) design review.* In the final design review, which is performed after the changes from the previous design reviews have been made, the product should be mature enough to enter into the final stage. The purpose of the final design review is to ensure that all of the requirements have been met. No individual should be held responsible for remembering all of the detailed information accumulated up to this point in a design. Rather, they should remember which details must be considered in the final design review.

If these design reviews have been properly documented and any involved requirements or changes have been identified, that should be part of the documentation package associated with each review. The most common errors evolving from such a review are errors of omission. Therefore, to ensure that a complete review of what is required is maintained, a series of checklists should be prepared and used to ensure that all review items have been covered. It is recommended that the checklists that have been included in MIL-STD-1521 be utilised. These checklists cover each of the specific design reviews in question for each and every programme element therein. Each design requires its own checklist, which should be carefully tailored to ensure that all of the people involved in the design review are fully aware of what it means.

In the case of reliability and maintainability, these checklists itemise detailed requirements for each of the functions. From the logistics point of view, the logistician should follow the results and the data output from each of the design reviews to ensure that it is properly inputted into the LSA so that the determining factors for each of the logistics element managers or supporting logistics elements are satisfactorily accomplished. Maintainability design requirements are the primary consideration in the final design review. The maintainability and reliability requirements, or other additional maintainability analysis and predictions and reliability estimates should be performed. Close correlation by reliability, maintainability and design personnel throughout the design phase is essential.

Continuity and follow-up

If the potential design improvement afforded by the design review programme is to be realised, continuity must be maintained from meeting to meeting. Also, recommendations must be followed up until the corrective action has been taken. Sufficient information must be carried over to succeeding reviews to avoid redundant coverage of similar problems. Continuity is difficult to achieve. It is essential that the documentation provides a degree of continuity, but it probably will not be sufficient to ensure adequate information transfer. Therefore, it is important that, as meetings are completed, everyone's ideas be documented and approved by all. This will assist in ensuring that continuity is maintained from meeting to meeting. One of the reasons for the generation of a final memorandum for the record is that continuity of personnel may not be possible.

The same personnel may not be available to repeat the design review assignments over an extended period of time, and they can seldom handle all of the levels of review. It may be possible for a permanent chairman to conduct all reviews on a given design, but it may not be possible to maintain the appropriate continuity between each of the people and all of the functions in question. Therefore, follow-up is necessary to ensure that the benefits actually accrue to and verify that appropriate design change actions have been taken, and that additional studies have validated the original design. When these two steps are completed and actions are closed, final documentation is prepared and validated by all of the people who participated in the original review.

Chapter 6 CONFIGURATION MANAGEMENT

Configuration management may be defined as a systematic approach that establishes and maintains the identification, control and accounts of the configuration of selected products and component parts.

Configuration itself is the physical nature of parts, assemblies, equipment or products, or any combination of these, that are capable of fulfilling the form, fit or functional requirements defined by the applicable formal specifications or engineering drawings. This is premised on the assumption that:

- the physical and chemical properties of the product that could be procured and/or fabricated are described in the specification;
- the performance of these products, as well as the functions required to achieve the performance, are also described by specifications;
- the product's fabrication or assembly to the specified envelope in conventional restraints is described by engineering drawings;
- the product's operational or maintenance concept is described and contracted.

To properly discuss the need for configuration management, one must realise that configuration management is a product of change. Without change, most projects could probably do without the services of an estimated 50% of the personnel who are not directly related to the manufacture, inspection and delivery of the products. It is therefore rather clear that configuration management is performed in one way or another by everyone who comes in contact with or is involved in the design, development and manufacture of an end item.

Configuration management should be carefully tailored to the quantity, size, scope and stage of the product life cycle and to the complexity of the configured item. The selection of items to be configuration managed is determined by the user's need to control an item's inherent characteristics or its interface with other items. The selection of prime and lower level configuration items (CIs) is basically a management decision normally accomplished through design engineering. The decision is based upon numerous engineering and logistics factors inherent in the specific design.

It is normally recognised that no single set of configuration management procedures will meet every need. This is due to the variations and requirements, organisations, industrial commodity areas and their working relations. The application specification should be tailored to recognise the particular programme requirements. The elements of configuration management are as follows:

- To assist in achieving, at the lowest possible life cycle cost, the required performance, realistic schedule, operational efficiency, logistics support and readiness of the configuration items.
- To allow the maximum degree of design and development latitude, yet introduce at the appropriate time the degree and depth of configuration control necessary for production and logistics support.
- To attain maximum efficiency in the management of engineering changes with respect to the necessity, cost, timing and implementation.
- Fundamentally, to ensure that
  - specifications, engineering drawings and valuable technical data are adequate for configuration needs and meet overall programme requirements;
  - configuration technical documentation is available when needed;
  - CI standardisation and compatibility are maintained;
  - total performance costs and scheduled impacts due to engineering change proposals (ECPs), deviations, and waivers are known at the time of their approval;
  - the CI's operational and non-operational use is known, and pertinent physical and functional interfaces among systems, equipment items and computer programs are documented and controlled.

Initiation of configuration management

The initiation of configuration management is based upon the final determination of what is required, how it will operate and under what conditions it will function. This comes together in the hardware and software specification. Once this document is generated and approval is granted either by the internal channels of the corporation developing the product, or by the developing activity, this becomes the departure point for the rest of the design. The acceptance of this document creates a CI. CIs can be generated during any phase of a programme's life cycle. Conceptual efforts need not be subject to configuration management. However, when applied in selected instances, the application should be limited to the functional characteristics.

Elements of configuration management

At any given time, configuration management can supply a current description of a developing hardware or software unit, product etc., and provide traceability to previous baseline configurations of that item. Configuration management also contains complete information on the rationale for configuration changes, thus permitting analysis and correction of deficiencies when they arise. Configuration management involves four functions:

1. Configuration identification.
2. Configuration control.
3. Configuration status accounting.
4. Configuration auditing.

An excellent description of configuration management practices can be found in US Air Force document AFSCP 800-7, *Configuration Management*. The following paragraphs are based upon that document, and help to explicitly define the four functions mentioned earlier.

Configuration identification and status accounting

Configuration identification is the family of specifications and drawings that describes the system of CIs during the design and development cycle. The identification becomes more precise as the design progresses towards production. This family of documents provides the basis for the development, production, testing, delivery, operation and maintenance throughout the total system's/CI's life cycle. CIs are identified through allocation of the systems specification requirements into lower tier requirements, which subsequently become prime item development specifications (PIDS).

Division of a product into CIs is a technical management decision. In other words, it is an acknowledgement that one item should be managed differently from another. Selection of CIs is a matter of judgement. Guidelines for CI selections are in the form of a checklist. The following questions should be used in selecting CIs tailored to the individual product in question:

- Is it a critical high risk, and/or a safety item?
- Is it readily identifiable with respect to size, shape and weight (hardware)?
- Is it newly developed?
- Does it incorporate new technologies?
- Does it have an interface with hardware or software developed under another contract?
- With respect to form, fit or function, does it interface with other configuration items whose configuration is controlled by other entities?
- Is there a requirement to know the exact configuration and status of changes to it during its life cycle?

If most of the answers to the above questions are 'no', then the item probably should not be a CI; if most of the answers are 'yes', the item probably should be a CI. If there are an approximately equal number of 'yes' and 'no' answers, then management must determine whether the item should be a CI.

Each CI should be produced by a single manufacturer and treated as a separate entity. The programme, or acquisition office should also limit the number of CIs in order to control the management effort and reduce costs to the procuring activity.

Once the CIs have been identified and the specifications defining them can be produced, the CI identification function must ensure that:

- all technical documentation describing the functional and physical characteristics of configuration management items is completely defined;
- verified technical documents defining the baselines are current, approved and available for use when necessary.

The CI identification number provides a permanent reference number for all CIs in a given type, model or series. Part numbers are usually needed down to the throw-away components of the lowest replaceable item.

Configuration status accounting is a management information system that provides traceability of configuration baselines and changes thereto and facilitates the effective implementation of changes. It consists of reports, records and documenting actions, that result from changes that affect the CI. Basic documentation includes the configuration identification index, which describes the approved configuration, and the configuration status accounting report, which describes the current configuration.

Configuration control

Changes to CIs can only be effected by a duly constituted configuration control board (CCB), as described in the following section. The CCB first determines a baseline which consists of the specifications that govern the CI design. Proposed changes to the design are classified as either Class I or Class II changes. Class I changes affect form, fit and function. However, other factors, such as cost or schedule, may also cause a Class I change. A non-exclusive list of potential items is given in Appendix B.

All other changes are Class II changes, examples of which include editorial changes in documentation and hardware changes such as material substitution, which do not qualify as Class I changes. Class I changes require approval by the procuring activity. Class II changes are identified as part of the configuration status accounting record, and are maintained by the manufacturer. These Class II changes are reported to the procuring activity so that there is constant awareness on their part as to what is being changed and why. Formal approval by the procuring activity is not necessary. Concurrence with the change is recommended.

Class I engineering change priorities will be assigned to each ECP based upon a selection from the following priority definitions. The priority will determine how fast the ECP is reviewed and evaluated, and how fast the change is ordered and implemented. The proposed priority is assigned by the originator and will stand unless the procuring activity has a valid reason for changing the processing rate.

Emergency priority. An emergency priority shall be assigned to an engineering change proposed for either of the following reasons:

- to effect a change in operational characteristics which, if not accomplished without delay, may seriously compromise national security;
- to correct a hazardous condition which may result in serious or fatal injury to personnel or in extensive damage or destruction of equipment.

A hazardous condition will usually require withdrawing the item from service temporarily, or suspension of the item operation, or discontinuance of further testing or development pending resolution of the condition.

Urgent priority. An urgent priority shall be assigned to an engineering change which is proposed to:

- effect a change in operational characteristics which, if not accomplished expeditiously, may seriously compromise the mission effectiveness of deployed equipment;
- correct a potentially hazardous condition which, if left uncorrected, could result in injury to personnel or damage to equipment. (A potentially hazardous condition compromises safety and embodies risk, but is within reasonable limits, permitting continued use of the affected

equipment provided the operator has been informed of the hazard and appropriate precautions have been defined and distributed to the user);

- meet significant contractual requirements (e.g. when lead time will necessitate slipping approved production, activation or construction schedules if the change were not incorporated);
- effect an interface which, if delayed, would cause a schedule slippage or increase cost;
- effect, through value engineering or other cost reduction efforts, lower costs to the procuring activity.

Routine priority. A routine priority shall be assigned to a proposed engineering change when an emergency or urgent priority is non-applicable.

Typically, a CCB meets weekly, but meetings may be convened on twenty-four-hour notice, according to the need. A typical configuration control process is shown in Fig. 6.1. It includes both the procuring activities and the manufacturer's functions. The process is initiated by any functional area within the procuring activity or the manufacturer's facility.

Before proceeding, a definition of what a CCB is and what it is intended to do is appropriate. This description generally applies to most manufacturers and procuring activities. The CCB is an organisation formed by the programme manager, who acts as chairman of the board. He is responsible for reviewing each of the proposed engineering changes submitted to the board by the various functions within the manufacturing operation, or those proposed by the activities associated with a procuring group. The board is normally called on a weekly basis, and the standing members of this board are:

- the Logistics Management group;
- the Engineering Management group;
- the engineer who initiated the original ECP;
- the Programme Management group;
- the Configuration Management group;
- the Manufacturing, Quality and Procurement activities.

While this is a standard group, alternative or additional people may be invited as required. Not all of these organisational functions need to be represented at all times. Their inputs are required to ensure that the appropriate design actions are performed and that necessary information is received by the programme manager so that appropriate decisions can be made. Each of these functional areas will review the proposed change to determine what impact it has on their relative function (see Fig. 6.2).

If a change that would have a major impact on the design is deemed unnecessary or too costly, or would make no major relevant change to the design, the programme manager will not authorise the suggested change. It is up to the programme manager and the functional area that originated the

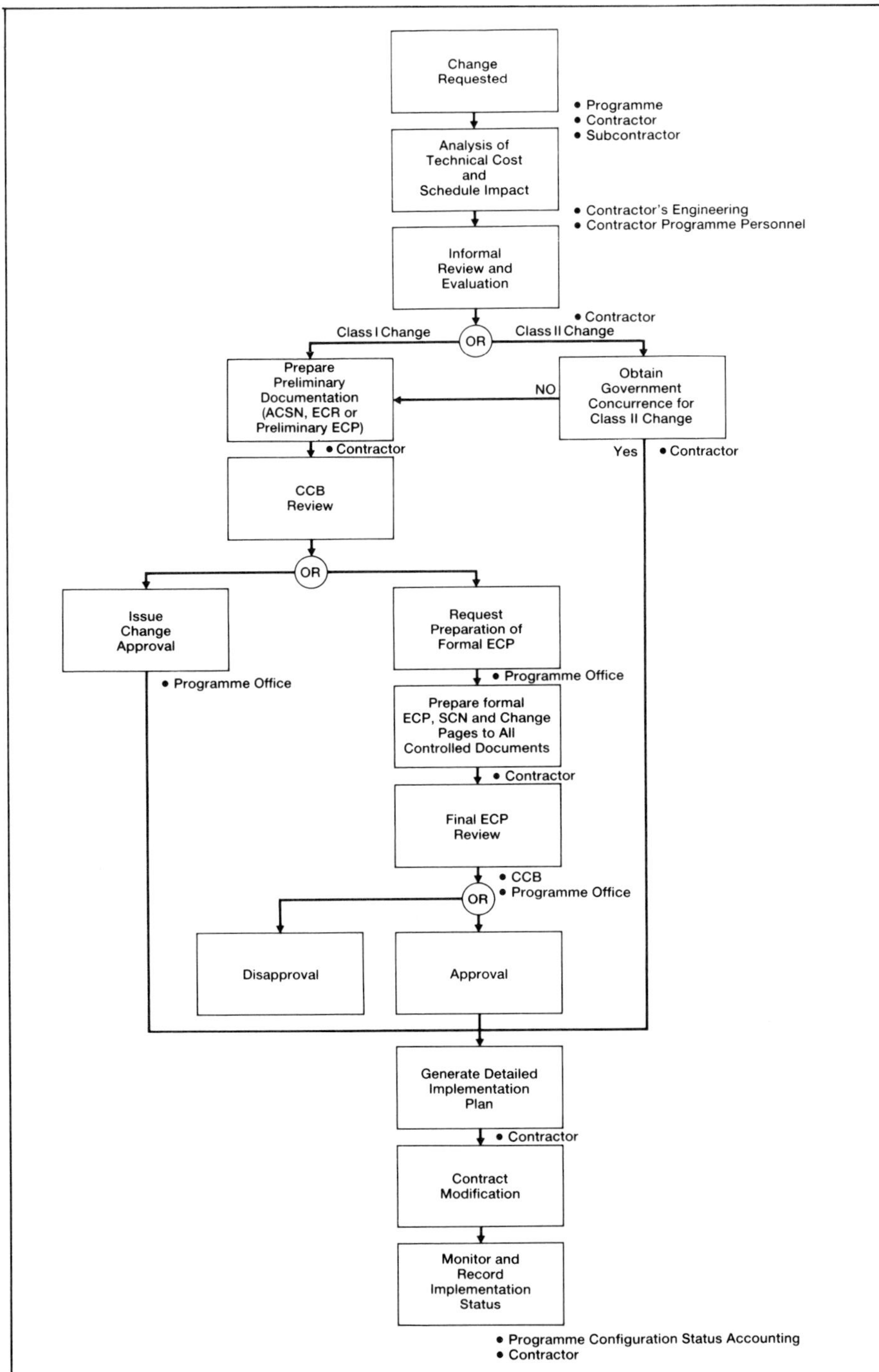

Fig. 6.1 Configuration control process. ACSN, advance change specification notice; CCB, configuration control board; ECP, engineering change proposals; ECR, engineering change request; SCN, software change notice.

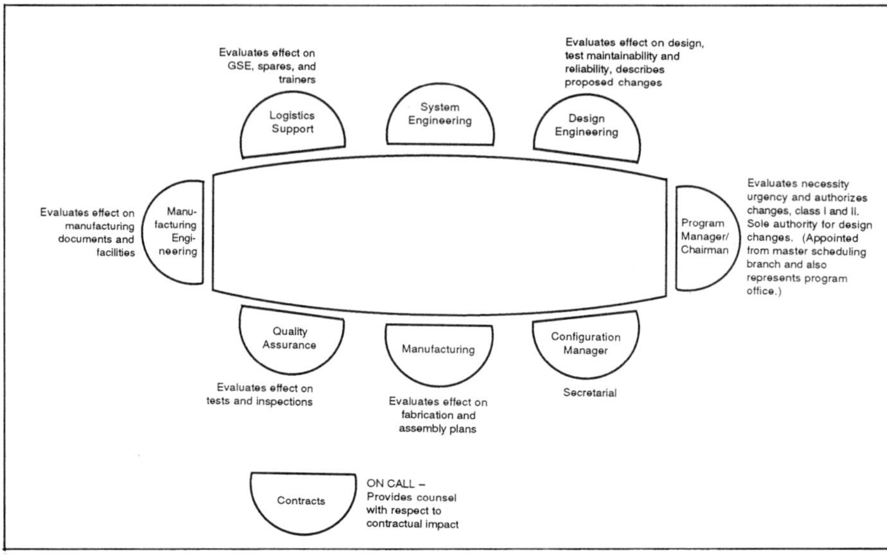

Fig. 6.2 Contractor's configuration control board.

proposed design change to determine what class the change falls into. If it is a Class I change, the programme manager has to make that determination and cost out what that change will be, determine its impact on the design and how long it will take to implement the design change.

This proposed change then becomes an engineering change proposal, which is forwarded to the procuring activity for their determination. This ECP must be approved by the procuring activity because it has a specific impact on the form, fit or function of the design. It is of importance to the procuring activity as well as the manufacturing facility to ensure that all relevant information has been included as part of the ECP (see Fig. 6.3).

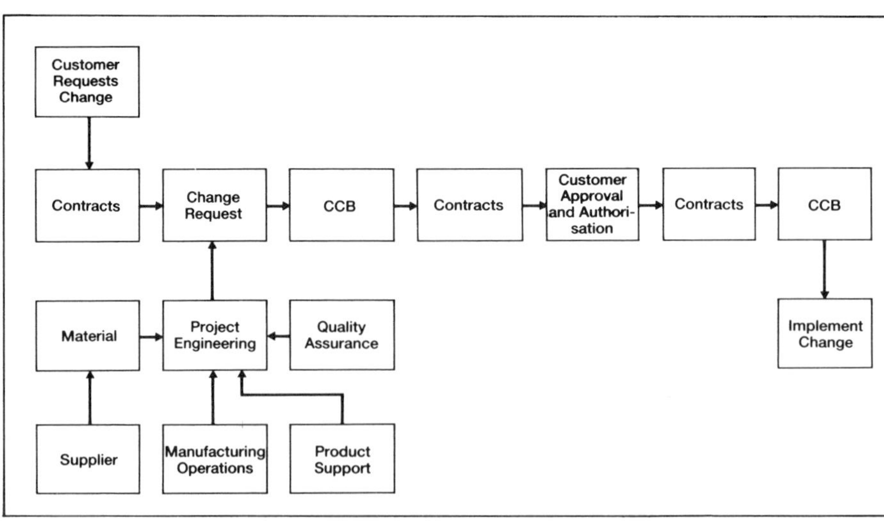

Fig. 6.3 Class I change control flow. CCB, configuration control board.

If the change has been approved by the procuring activity and requires additional funding, this must be done by contract modification. If no additional funding is required and the change falls within the general scope of the requirements of the contract, the approved ECP is forwarded to the manufacturer, who is responsible for integrating the change into the design. The timing of this type of change to the design will have an impact on the actual schedule and development of the product in question. Rapid response to change is essential. It is done to ensure that the cost to the programme is kept to a minimum, and that all of the documentation associated with the hardware will be changed to correspond to the proposed engineering change notice (ECN), which will be generated to support the ECP. A written record of all actions taken by the CCB will be kept by the secretary, who is the configuration manager, and maintained as part of the total configuration audit and status accounting report. The actual change then becomes part of the contract, and must be traceable back to the original baseline. Fig. 6.4 shows the impact that a change will have on a total system. This figure represents changes that occur during the production phases, but is also applicable to changes that occur early in the design phase, specifically during the FSED phase of the procurement life cycle.

Configuration audits

Audits are used to validate that the developmental requirements have been achieved and that the product's configuration is identified. This is accomplished by comparing the configuration of the CI with its technical docu-

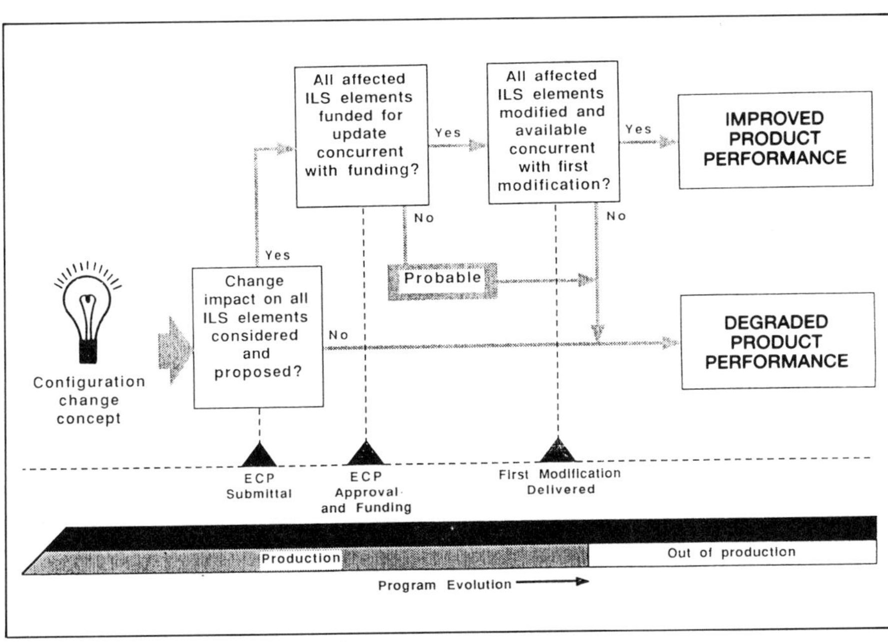

Fig. 6.4 The effect of change on the system.

mentation. Two kinds of audits are performed: functional configuration audits and physical configuration audits.

The functional configuration audit (FCA) is a means of validating that the development of a configuration item has been completed satisfactorily and that the item functions as required. It is a prerequisite to the physical configuration audit. A functional configuration audit is normally performed during the FSED phase, just prior to production.

The physical configuration audit (PCA) is a means of establishing the product baseline as it is reflected in the production configuration identification. It is used for the production and acceptance of units of configuration items. The functional configuration audit may be accomplished during the FSED; however, it is usually delayed until the beginning of the production phase so that it can be accomplished on an early representative production unit. A physical configuration audit is normally required on the first CI to be delivered by a contractor.

Chapter 7 INTEGRATED LOGISTICS SUPPORT: THE PRINCIPAL ELEMENTS

This chapter will cover the principal elements associated with the development of a proper integrated logistics support programme. Each subsection will be established within this chapter so that the reader will have a better appreciation of how all of the elements interrelate with one another. Each paragraph will describe in detail the specific element, how it works and what its outputs are. The subjects to be covered have not been prioritised. This approach makes it easier for the reader to appreciate how, when and why each element is formed.

Maintenance concept and planning techniques

Background

To prevent haphazard approaches to maintainability design, a definitive concept or plan of product maintenance is essential. Such a plan would define the repair philosophy to be used, guide support planning and provide a basis for defining support requirements. In establishing a plan, two broad objectives should be kept in mind:

1. Optimise each level of maintenance consistent with cost and with the system operation requirements.
2. Minimise the requirement for highly skilled personnel.

Formulation of the maintenance concept demands a thorough analysis of the product structure, function and use in the precise definition of repair policies. This concept is the principal factor in formulating the maintainability requirement. It is also the impetus from which all other logistics elements are derived. If, for example, the repair philosophy is to replace modules, modularisation must be incorporated as a basic element of product design. To be effective as a device for formulating basic design characteristics, the maintenance concept must be fully translated into maintainability requirements. In particular, the repair policy, if incorporated, must be clearly stated. The planners must give careful consideration to the applicability of all support levels.

To successfully identify the basic objective of a maintenance planning concept or a maintenance programme, the concept is developed to be consistent with the operational requirements of the proposed product at the lowest possible cost. The concept is developed as soon as possible and in conjunction with the definition of operational requirements. During subsequent phases of the procurement, the maintenance concept is updated to consider practical trade-offs for risk avoidance between the operating requirements and the design engineer. Reliability and maintainability are key considerations in the acquisition, and every effort should be made to optimise maintenance and total product costs.

The concept is developed for maintenance support and is tailored to the basic structure of the evolving product. This ensures the most effective and efficient means of maintaining resources and sustaining the required degree of operational readiness. The maintenance concept delineates the approaches that should be taken at the various levels of maintenance. It establishes criteria used by the design engineer to perform the trade-off studies. These studies are used to compare the different types of maintenance techniques (whether automated, semi-automated or manual) that are necessary to successfully meet the quantitative values set forth by the maintainability engineer.

As discussed in the previous chapters, the maintenance concept is a principal factor in determining logistics support requirements. It is supplemented by the logistics support analysis, which leads to the identification of the maintenance tasks, the task frequencies and timing and the required number of personnel to support each level of maintenance and their levels, along with the test and support equipment necessary to repair and replace failed units.

Maintenance concept alternatives

Formulating the maintenance concept and making the discard or repair decision are highly dependent activities that should be accomplished in parallel as the product evolves. One of the problems that we are beginning to face nowadays with the highly complex products that are being designed and developed is the ability to identify early in the programme the amount of computer memory that will be required to establish the appropriate diagnostics necessary to troubleshoot the product once it is deployed. The ability to identify which circuits must be tested, how frequently the testing should be done and the degree of impact the software programs would have on the entire operation are important factors that must be considered by the design engineer and the maintainability engineer as early in the programme as possible. Without this teamwork, it will be highly unlikely that the maintainability engineer will be able to include the necessary system diagnostic routines in the computer program because the design will more than likely grow and require as much computer memory as possible. Therefore, it is essential that when the reliability and maintainability requirements are defined, the diagnostics required to support the programme also be defined.

The approaches by which the diagnostics are applied will depend upon the various approaches and alternatives available to both the maintainability engineer and the designer as the equipment begins to take shape and the functional partitioning begins to evolve. The maintainability engineer should examine each feasible maintenance alternative and combination of alternatives from the aspect of both cost and operational requirements. Fig. 7.1 represents all of the perceived maintenance alternatives that the maintainability engineer should consider. The various alphanumerics on the figure will assist in the identification of any particular alternative. For example, assume that 1, 2 and 3 represent the moving and replacement of an assembly. The assembly is sent to the intermediate level of maintenance and is repaired by part replacement and returned to the supply. The replaced part is then discarded. It is apparent that many alternatives are available for consideration. Fortunately, most of them can be eliminated by system characteristics and operational requirements. The application and use of diagnostics will be directly related to the speed with which the requirements have identified the maintenance technician to bring the equipment back into operation.

The diagnostics developed for the product will be used to help isolate and localise specific problems. Assuming, for argument's sake, that there is sufficient money, time and computer memory, diagnostic routines can be devised to check approximately 99% of all active parts within a product. This becomes very costly if the problems are not properly identified and the diagnostics are not properly applied. This becomes a very important aspect of the total system design, because if it is not considered early in the programme, the ability to place the electronic test points will be hampered and the maintenance technicians will not be able to identify where the problem

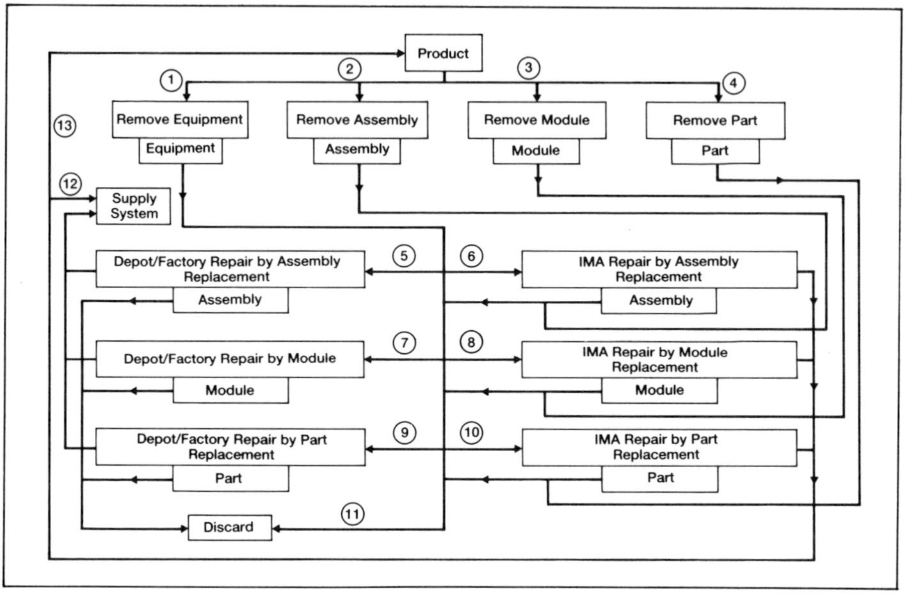

Fig. 7.1 Maintenance alternatives. IMA, intermediate maintenance activity.

exists. The use of diagnostics only permits the identification that there is a problem within the product and the isolation of the problem to a specific area.

The maintainability engineer must play a dynamic role in the design programme so that he can assist the design engineer in physically locating the equipment, unit or assembly in their appropriate location, thus making it easier for the maintenance technician to remove and replace the failed component. As an example, no maintenance concept beginning with alternative 'A' would be practical for a shipboard subsystem, because the equipment in that platform would be far too large to be removed and replaced. On the other hand, it may be quite feasible to employ a maintenance concept such as that described by 11 in Fig. 7.1 (remove/discard) for much simpler, low cost products. The maintenance alternative is not selected on the basis of cost alone. There may be trade-off decisions regarding reliability, maintain-ability and other elements of effectiveness. The proper procedure for selecting the maintenance alternative depends upon the platform, equipment or system in question. It must be tailored to satisfy the specific needs and specific requirements as defined in the development specification.

To help determine what the maintenance concept should be, level of repair analysis has been developed to assist the maintainability engineer in determining the lowest levels of repair actions that should take place for each equipment, assembly or component in question. The level of repair analysis will assist the maintainability engineer in determining the appropriate level (whether it be organisational, intermediate or depot maintenance) at which the part should be repaired. To ensure recognition of the varying options that can be made regarding the maintenance concept or the system diagnostic test approach, alternatives should be limited in specific ways. The testing categories that should be considered are: (a) no testing, (b) internal and external manual testing and (c) internal and external semi-automatic and automatic testing.

These terms imply the following:

- *No testing.* Discarding equipment or systems upon failure. This will depend upon the size and cost of the system being tested and whether it is economically sound to discard rather than repair it.
- *Internal testing.* The use of built-in test equipment to detect, localise and isolate failures.
- *External testing.* The use of external test equipment to detect, localise and isolate specific areas in which a failure has occurred.
- *Automated testing.* The use of methods that will detect, localise and isolate failures without the attention of the maintenance personnel. This will depend upon the amount of computer memory available in the system being developed. As a system becomes larger and more complex, more computer memory may be made available to the maintenance engineer so that his application will simplify the maintenance technician's future role.
- *Semi-automatic testing.* The use of methods that automatically perform procedures for detecting, localising and isolating failures, but require maintenance personnel attention to go from step to step within the procedure.
- *Manual testing.* The use of methods that require maintenance personnel capable of detecting, localising and isolating failures, primarily by trial and error techniques utilising available standard test equipment. This approach is normally used when there is limited funding available to institutionalise an automatic approach. It also denotes the fact that there may not be a computer designed into the equipment to help the maintenance personnel establish the alternative capabilities necessary to troubleshoot the system. With the techniques available, and with computers now becoming smaller and smaller, the abilities to automatically or semi-automatically test the equipment in question are becoming more common, thereby eliminating the need for manual testing.

Table 7.1 shows the maintenance plans with respect to the total maintenance time required to detect, localise and isolate the specific failures. Each of these approaches has a direct impact on what the logistics engineer will need to have on hand at the time of a failure and how difficult it will be for the logistics engineer to define the requirements (such as spare parts, training, technical manuals and so on) to support the product.

The following paragraphs summarise what each of the above terms means and the impact they have on logistics.

Table 7.1 Relative maintenance time for different maintenance plans

Maintenance time	*Maintenance plan*
Longest	External manual testing Internal manual testing External semi-automatic testing External automatic testing Internal semi-automatic testing Internal automatic testing
Shortest	No testing

No testing. The only factors other than maintenance time to be considered for the 'no testing' category are cost, total cost of discard and logistics.

- *Cost.* Is the product or its components inexpensive enough to be discarded upon failure?
- *Logistics.* Will the supply system be such that replacement equipment or products will be available for use when failures occur?

Automatic, semi-automatic and manual testing. Factors other than maintenance time to be considered in selecting from these categories are as follows:

- development time for test equipment;
- test equipment costs;
- operational planning and deployment of end item;
- amount of testing to be performed;
- readiness requirements of the prime equipment;
- maintenance echelons involved: organisational, intermediate or depot;
- simplicity or complexity of equipment in question;
- training and training costs; availability of personnel.

The use of automatic or semi-automatic diagnostics should be considered only when one or more of the following conditions exist:

- turnaround time and down time must be held to an absolute minimum;
- many repetitive measurements must be made;
- readiness requirements dictate that automatic or semi-automatic testing should be used;
- maintenance loads are heavy.

The choice of automatic versus semi-automatic system diagnostics requires answers to the following questions:

- How should the test equipment be programmed?
- How should test results be displayed: go/no go lights, meters, colour codes, readouts, etc?
- Should testing be stopped when an out-of-tolerance condition is detected, or should the testing branch automatically into isolated routines?

The questions should be constantly reviewed and evaluated as the design matures. The ability to integrate all of the diagnostic routines will depend on the amount of memory available and the cost necessary to permit the design of diagnostics to troubleshoot each of the major functions of the product. Diagnostics routines should follow the failure modes and critical effects analysis results. It is important that the critical elements of the specific design are checked so that the elements that appear to have major impacts on the product's operation are highlighted.

The other aspect to be considered besides the computer memory and cost factors is the reliability factor associated with each of the items in question. Equipment that tends to show high failure rates should be tested more often than equipment that indicates low failure rates. The amount and degree of testing will be a judgement call on the part of the maintenance engineer and supported by the reliability and design engineers. The sequence of testing should always be based upon checking the tester first. Then, in a serial pattern, test each individual box, from the front end to the tail end of the product. By performing the tests in this sequence, it is ensured that the signals are being processed in proper sequence, with all of the appropriate signals being given to the appropriate areas.

Depending upon the amount of storage available in the computer or money associated with supporting the maintainability and establishment of diagnostics, two steps would need to be reviewed. First, would the design and maintenance engineers automatically allow the system to start diagnosing the equipment that it recognises as having failed? This is done by placing test points between the major units and then, with more defined subroutines, checking out each unit within the equipment level so that the equipment can be troubleshooted to the lowest replaceable assembly. The other option available to the maintenance and design engineer is to determine which unit has failed and pass the problem down to the lowest replaceable assembly. This can be accomplished through a combination of tools, human interface, or predesignated and located test points. Built-in test equipment or furthering the use and application of diagnostics is also applicable.

The determination of what the lowest replaceable assembly would be is based upon an earlier analysis performed by the design engineer and the logistician/maintenance engineer, via the logistics support analysis (level of repair analysis). These costs will assist the design engineer and the maintenance engineer in determining the most cost effective approach to deal with a possible problem. If the determination is to localise the problem using human interface and standard test equipment, the design engineer and the logistician should place test points in areas that are easily accessible to the maintenance engineer. The number of test points should be related to the number of replaceable units and the number of active circuits in each of the replaceable assembly. A test point is generally required for each output from and each input to every unit that is replaceable or the specific maintenance echelon being considered. Test points may be either the signal sensing type or signal injection type. In addition, if adjustment facilities are provided on a given assembly, a test point must be provided that will permit direct observation of the effect of each of the adjustments.

These requirements must be considered for each level of maintenance to establish the total number of test points required. Practically all electronic circuitry follows one of three basic data flow patterns. Individual system elements are arranged in a recognisable flow that becomes the basis for determining the replacement of accessible test points. Fig. 7.2 presents these flow patterns as block diagrams. This figure has been prepared for use in determining test points using systematic diagnostic approaches or standard test equipment. The elements are not identified, since each box can represent any element of the product, depending upon the maintenance echelon being considered. The test principle remains the same, whether the element is a discrete part, a module or an assembly.

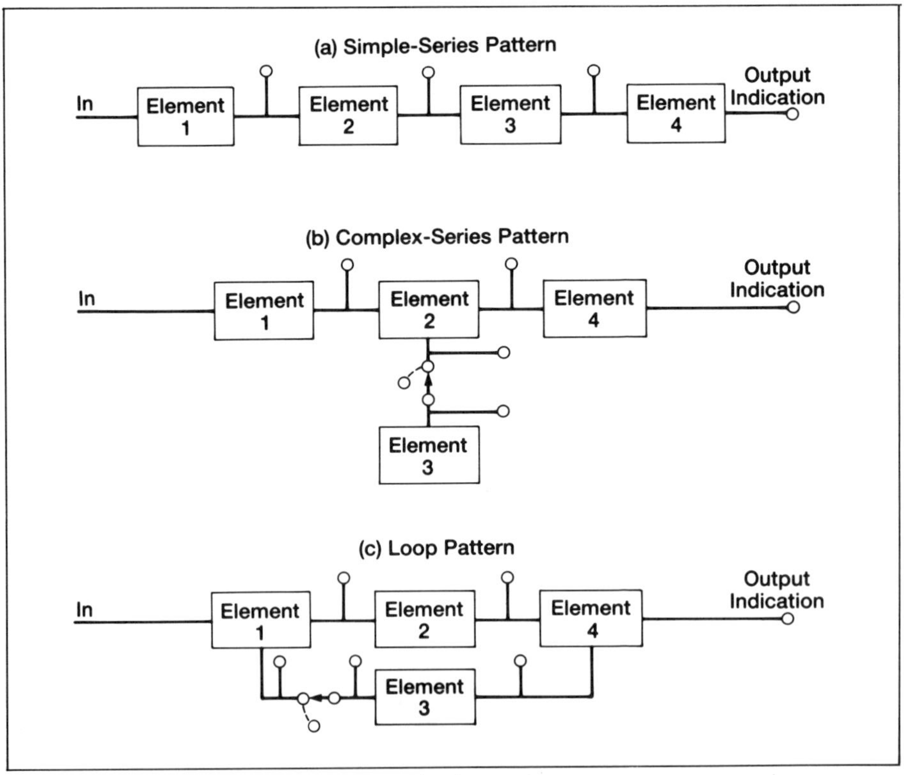

Fig. 7.2 Data flow patterns encountered in electronic systems.

The simple series pattern represents the simplest pattern for test point location. It can also be recognised that this is the highest level of diagnostic testing, and it includes the ability to identify which element, 1, 2, 3 or 4, has actually failed in the sequence of events. The output represents the failure of an unknown link in a functional chain, but the chain is well defined and each element performs an operation on a single output. Fault isolation consists of testing along the pattern until the fault element is detected. For the simple series presented in Fig. 7.2, assuming a good input, three access points located as shown will provide isolation to the element level.

Locating access points for testing a complex series pattern involves greater sophistication. Failure indication may depend upon the position of a switch at the time of the failure. In the simple series pattern, the chain is defined, but with the complex series pattern there are two potential flow definitions, and the access points must be located to observe both. Four access points are required for fault isolation.

The loop pattern is the most difficult for fault isolation, since the indication for some elements depends on the proper operation of others. Therefore, the indications may not truly reflect the state of the element, but the effect of the failure of a related element. Fig. 7.2 shows that at least five access points are required to fault isolate to the element level, and that provisions may have to be made to open the loop and evaluate test results for more than one circuit condition.

Physical considerations

One of the most important aspects of establishing locations of test points for future use by the maintenance technician is the accessibility of the test points from both a testing and a visual standpoint. The physical location of the test points depends upon the testing method selected (primarily, whether built-in testing is being used). When external test equipment is being used, accessibility and convenience are major considerations. When internal, automatic or semi-automatic testing is being used, these factors are of minor importance. Instead of the sensor size and type, the requirements of signal convertors and the efforts of lead length dictate and indicate the physical location of the specific test points. When external test equipment and manual testing are used, test points should be suitably labelled so that they are physically located close to the replaceable item being tested. Thus, failure indications at the test points can be associated with a specific assembly to be replaced. Specific types of test points that could be used are:

- red lamps that are lit when a failure occurs within a specific circuit or when the failure of a circuit initiates a circuit break;
- bypass signalling that indicates that a circuit has failed; or
- opening of a switch.

Any combination of these devices can be used to assist the maintenance technician in troubleshooting the assembly. The level of complexity of the equipment in question will help the design engineer to determine the application and implementation of the various approaches. The time required to bring the product back up and make it operational will be the driving force behind this entire technique. This is directly related to the maintainability engineer's needs for achieving the specific requirements identified in the specification. Detailed information on identifying test point accessibility, using either manual testing or automatic and semi-automatic testing, is inappropriate for inclusion here. Specific handbooks and other detailed information are available to the maintenance and design engineer that would provide much more information than could be included in this book.

It is very important that the maintainability and maintenance engineers and the human factors engineer work closely together, because what the maintenance engineer suggests to the design engineer may have a direct effect on the human engineering efforts associated with a specific design. Therefore, all team members should be closely coordinated when designing techniques for implementing a system.

Maintenance planning

With the formulation of the maintenance concept, the establishment of diagnostics routines and appropriate test points, and the location of the equipment necessary to support the maintainability engineering design approach, the logistics engineer should now establish the foundation for the maintenance plan.

The maintenance plan is normally generated toward the latter part of the FSED or early in the production phase. The plan specifically identifies and includes a description of all recommended levels of maintenance, the responsibilities of the maintenance technician and the applications and results of the level-of-repair analysis, so that the maintenance planning document will identify which part is sent where, and what level of maintenance will be applied once it arrives at the consumer's location.

The maintenance plan, therefore, is a culmination of all of the work performed by the various functions involved in logistics support and design engineering. One must recognise that the maintenance plan is an output of the efforts of the maintainability engineer, the maintenance engineer, the logistician and the design engineer. The foundation for all of the logistics activities, though, is based upon the efforts put into the generation of the maintenance concept. This is specifically defined and detailed as part of the maintenance plan so that the operating and maintenance personnel will have an appreciation of how the product is going to be checked and to what level the checking is performed.

However, one must recognise that the maintenance planning document is an iterative document. Even though it is generated towards the final stages of the design, the LSA work that went into the generation of this document will constantly be reviewed. Changes to the LSA automatically create changes to the maintenance planning document; therefore, the logistics engineer must carefully monitor the timeliness of each of the changes in the LSA so that, prior to the distribution of the document, the latest information is available to the people who will be responsible for maintaining the product after it has been distributed.

Appendix C was prepared to give the reader insight into what a typical maintenance planning document should include. The proposed approach is all-encompassing, and should the reader need to generate such a document, it would pay him to carefully examine the appendix to determine what should be included as part of the maintenance planning document.

Conclusions

The logistics engineer must recognise that the maintenance concept is the most critical element in the entire logistics inventory. Every one of the logistics support elements is directly affected by what occurs with the maintenance concept. Fig. 7.3 portrays that impact. A different way of describing it is illustrated in part of Fig. 7.4. This demonstrates that, as a result of the maintenance concept and with inputs from the reliability and maintainability engineering efforts and the training department (which establishes personnel skills), the maintenance concept drove the diagnostics, which eventually established the test point requirements. From the diagnostics derived from the maintenance concept, the training, technical tools, test

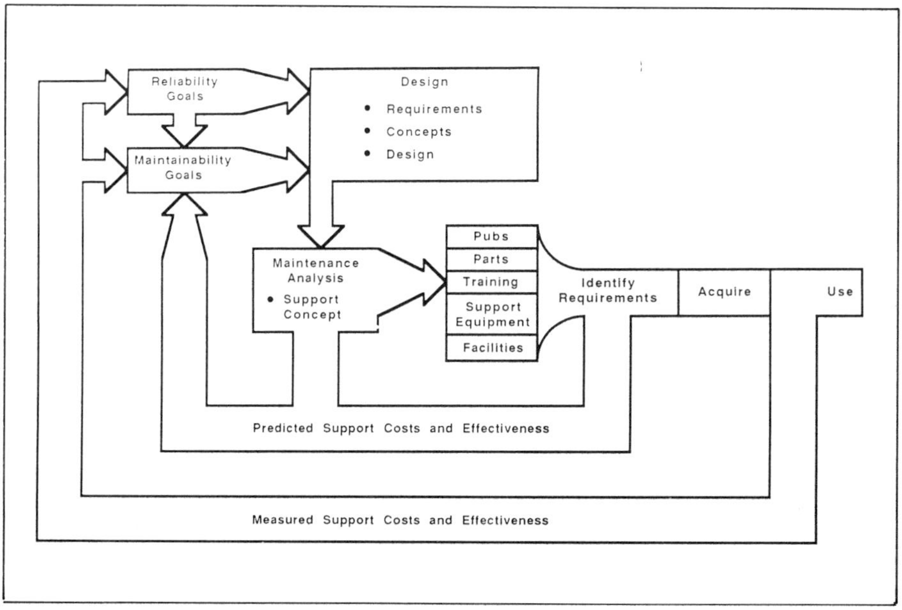

Fig. 7.3 ILS functional relationships.

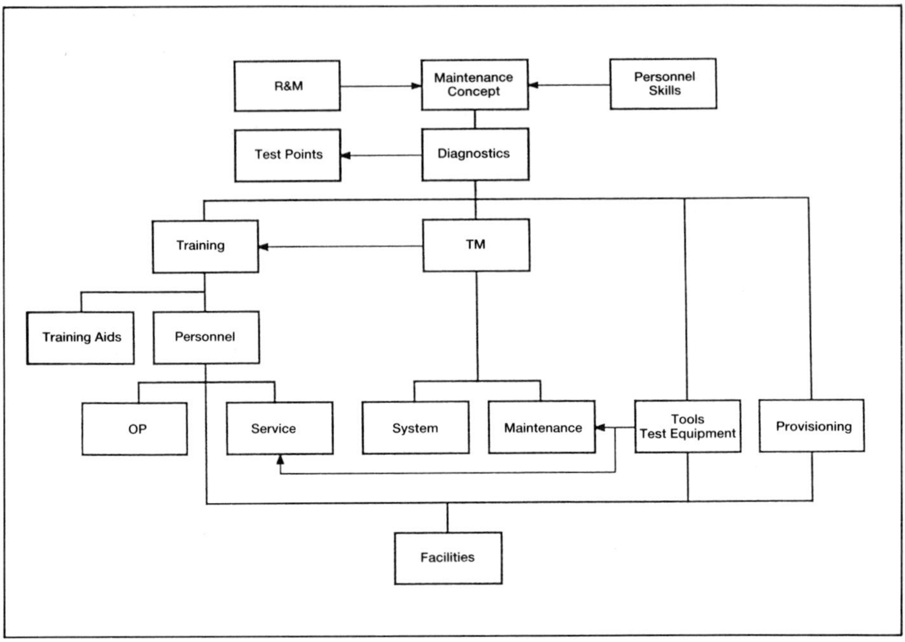

Fig. 7.4 Maintenance planning hierarchy. OP, operations; R&M, reliability and maintainability; TM, technical manuals.

equipment and the provisioning needs will be defined. All of this is discussed further in Chapter 8 on the application and use of the logistics support analysis.

The importance of maintenance concepts and planning cannot be overemphasised. Logisticians must recognise that if anything is to occur in accordance with the optimisation of logistics and if they are to ensure that all of the supporting functions are properly defined and activated, the maintenance concept must be defined so that it minimises the impacts on all follow-on logistics efforts. The mechanism through which the application and use of the maintenance concept is measured is through the logistics use of life cycle costing. That is the input to the logistics support analysis, and it is used as part of the total package in defining what the final system configuration will look like.

To reaffirm this concept, the maintenance concept evolved from the definition of the product's operational requirement, and delineates:

- the anticipated levels of maintenance support;
- general overall repair policies and/or constraints;
- the organisational responsibilities for maintenance;
- major elements of logistics support as they are defined for a new system;
- the effectiveness requirements associated with the system support capabilities;
- the maintenance environment.

The maintenance concept basically describes, in general terms, the overall product support environment and constitutes the baseline for the determination of specific logistics support requirements through the logistics support analysis.

Manpower and personnel

Personnel and skills

All recognise that it takes people with the appropriate skills to operate any new piece of equipment or product that is either distributed or fielded in the commercial or military environment. The first question to be asked is, what skill levels will be used with the specific pieces of equipment or products. From a military standpoint, the complexity of the military equipment is becoming much greater and is forcing an increase in the skill levels of the personnel that are available in the services. In today's world it is recognised that this is becoming less feasible because the people entering the military now are not as well educated as those in previous years. Therefore, the ability to determine the skill levels of personnel is an essential element prior to and during the development of any new product.

From the commercial point of view two aspects must be examined. The first is, can the consumer use the equipment and is it easily understandable by the person who is going to use it (how to turn it on, how to operate it and how to turn it off). The second aspect is, once it fails, is it replaceable or repairable, and will a duly described, accepted or certified technician be capable of repairing the device as prescribed by the documentation. These questions, either from the military or from the commercial standpoint, must be quickly answered before any new product reaches the drafting board. Before describing the skills of the required personnel, let us take a good look at (1) where these skill levels come from, (2) what is required to ensure that the skills of the personnel who will be using it are sufficient and (3) how many people will be required to support the product.

Skills and skill levels – where they come from

From a military standpoint, the personnel who are already in the military services have already acquired a certain level of skill. They have gained experience through actual operational use of similar pieces of equipment. This does not necessarily mean that they have kept current with the advances in technology. What it does mean is that those people are already available and have some basic knowledge of how to operate electrical, electro-mechanical, or mechanical equipment. The question really comes down to how much additional training and expertise is needed to ensure that the product can be supported by the people who are already in the field.

In today's environment, the level of experienced personnel is decreasing. Those in higher levels, such as non-commissioned officers are not staying in the Government as long as they had been previously. The up and coming cadre of personnel will take longer to achieve higher levels of expertise, and they turn over more quickly than those in the past due to the enticements to return to the private sector. Now the question is, what should be done to ensure that the personnel that are already in the military service will have the proper capabilities and knowledge base to be able to perform the tasks at hand? This will require the training of additional personnel (see pp. 128-133). The equipment designer should go back to the acquisition manager or the procuring activity to determine the types of personnel that will be made available to support the programme. It should be recognised that no additional programme or support personnel will be added to the programme, therefore we have to use existing personnel. Once the material, skill level and knowledge have been determined, the designer and the systems engineer should determine the tools which could be designed into the equipment to simplify the work to be performed by either the skilled or unskilled worker available in the military. The first and foremost consideration is that the enlisted personnel know how to read. In the US military, the reading skills or the educational level of the unskilled personnel average at this time frequently no higher than the 7th grade. The capabilities of these personnel to quickly assess and determine what is wrong with a piece of equipment decrease with the lack of educational capabilities.

Therefore, it is necessary for the designer to automate as much of the equipment as possible, and allow as little of this information to flow back to the operator. As a result, all he has to do is read a simple set of instructions so that he can follow more quickly than ever before. This is a must. Even though this tends to complicate matters, it could simplify the process in the long run. The advent of the computer, the increased speeds of the computer capability and the new languages being introduced provide the tools by which all of this can occur. The personnel and the skill levels are tied directly with the maintenance concept. They are also tied to the operational idea that is being proposed as part of the overall contractual requirements.

Needs of personnel and people

Needs of personnel and people are essential from the standpoint of the designer, because if there is an insufficient number of people to do what the designer intends, then a basic problem exists in being able to operate or to maintain the equipment. There are two specific areas concerning manpower that we must consider during the design phase.

The first is the number and type of maintenance personnel. The second, which is equally important, is the skill levels of the individuals responsible for operating and manning the equipment once it has been deployed. This becomes an extremely touchy effort; unless the new product will replace an existing one, the use of more manpower than was stipulated by the original contract will be difficult and time consuming. The procuring activity must fill the billets that will be required to support the fielded product. The logistician must try to quickly assess what the new skill level requirements will be, and how much additional education will be required. This is to ensure that once these people are brought into the operational environment, they will quickly learn the equipment and be able to operate it as designed. Such an assessment will help to determine whether to modify the design early in order to support exactly what these personnel are supposed to accomplish.

As mentioned before, the maintenance manpower requirements are extremely limited and at the present time they have been stretched to the ultimate degree. The maintenance personnel can only do so much in the time they are allotted to work on the problems. The more problems they experience, or the worse the reliability of a product becomes, the harder it is for them to ensure that the equipment will be properly maintained and kept in operation. Therefore, the reliability of the programme becomes an essential element and attempts to help maintain or reduce the current requirements to support this new product. Therefore, the impetus for personnel and manpower becomes reliability, the maintenance concept and designs of the diagnostic routines as part of the maintenance concept. The more the equipment can do to identify its specific problems, the easier it will be for the maintenance personnel to repair the device.

All of these issues must be resolved early in the design phase. The design engineer and the logistician should, as time permits, require of the recipient regarding the following items:

- the skills levels associated with the people currently working on the job for which the equipment will be required;
- the detailed knowledge bases from which these people are coming;
- what schools are presently available within the military to help them gather the appropriate information and knowledge base.

In this way, when the new product is distributed it will be a lot simpler for the manufacturer to operate or maintain than to wait until the equipment is deployed and then try to gather the sufficient information from the customer to support it. This effort of data gathering should be done as early as possible. The training aspects of the programme become an integral part of the total acceptance of the equipment. The present skill levels associated with and cooperatively supported by the training programme will be an asset to the firm. The two together will improve the personnel capabilities and give the corporation the necessary image, because now the operating personnel can satisfactorily perform the tasks at hand.

The commercial side is just as difficult. This is because one is not sure what the recipient of the equipment will know and what his or her capabilities are. The important facet of this is to know the consumer: whether it is the everyday person or experienced personnel who are responsible for factory maintenance engineering. The surveys that should be performed to assess this are absolutely critical to a commercial product. Consumers nowadays are demanding a lot more for a lot less. The problem lies in their ability to recognise when they have a problem and how best to fix it. There are too many tinkerers in today's world. Therefore, it is necessary to be able to identify quickly whether there is a problem with the product, and whether that product can be easily fixed by the consumer. If it is not easily fixed, or the consumer does not wish to get involved in making the repairs, then the distribution or the repair centre must be made available to him within a reasonable period of time so that these items can be fixed. This raises a second question: will the technician in the commercial environment be able to fix these new electronic devices? If not, what will it take to train these people to make that happen? Additional training will be required but, again, the manufacturer or the logistician should know what the skill levels of these personnel should be. The best way to find this out is through the field service engineers. They are the people who are specifically trained to interface with the manufacturing support persons, find out what they do and do not know, and set up the appropriate training programmes to satisfy their specific needs.

Skill levels vary with people. This means that their knowledge bases vary. Therefore, the design of the product should be such that they can easily replace what is required to be replaced and ship that part back to the original manufacturer. This also creates a problem, because it has an impact on (1) the distribution and turnaround time, (2) the transportation, (3) the inventory and warehousing requirements. All of this impacts the logistics support cycle associated with the product. If the equipment is properly designed to meet the individual human capabilities, then the logistics chain from a manpower and personnel point of view is reduced.

The balance between the training skills and the product design should be constantly reviewed, because everyone does not have the same or appropriate skills to do a complex job. Regardless of the working environment, military or commercial sector, the skill level is the basic criterion from which everything is generated. The lower the skill level, the more complex the device becomes, because now the device must tell the operator what is wrong with it. The higher the skill level, the less complex the device becomes as regards diagnostics.

If properly used, the information tendered during the design phase will become a source of data identified as part of the logistics support analysis techniques. There are specific elements associated with the necessary information to identify the manpower and personnel requirements. Regardless of what business we are in, military or commercial, the application and use of LSA will assist the logistician in determining what manpower and personnel will be required to support the tasks at hand. One of the basic precepts in the logistics environment is to train the people properly, and have the right skill levels at the right place, at the right time, doing the appropriate job, because this has a direct impact on the corporation. The personnel and manpower that will be doing the job will be the manufacturer's link into the service sector. The support people will be interfacing with their supervisors, relating how successful the training was.

Logisticians should ensure that the appropriate skills are at hand and that necessary training has been included in these people's curriculum so that the manufacturer can limit the amount of negative exposure if the training personnel are not successful in completing their job. This is a manufacturer's first and last line of defence. The affirmative position is that the offensive moves made by the manufacturer ensure that the people handling his products are properly trained, and the necessary support is provided to ensure that they can successfully complete the work they are being asked to perform.

System supply support

Supply support consists of all management actions necessary to determine the acquisition cataloguing, packaging, preservation, receiving, storage, transfer and issue and disposal of spare parts, repair parts, consumables and special parts needed to support both scheduled and unscheduled maintenance.

Supply support is therefore the process through which the required system support is made available to the user. To achieve these initial objectives certain documentation must also be generated so that the LSA can be provided in the appropriate format. Such inputs are normally generated through the use of reliability predictions and modelling techniques, the results of test data (if and when applicable), field data as appropriate and historical data. Also, the preliminary maintenance analysis and equipment plan should be generated so that sufficient information will become available. All of this information becomes part of the LSA process, and the reports for provisioning are output via LSA report 036.

The maintenance concept discussed on page 101 is a primary consideration for the resulting needs of the supply or spare parts requirements listing.

Within the parameters of the supply support concept and during the selection process, piece-part versus modular replacement can be considered. Factors such as the cost and feasibility of increased reliability versus reduced quantities of repair parts and the effect of human engineering design on maintainability should not be overlooked.

The concepts formed at this stage will have a direct impact on other items later in the product's development cycle. Items like equipment publications, training programmes, facility requirements, stockage locations and operational capability will be affected.

Therefore, everything that precedes the generation of provisioning technical documentation will need to be integrated into the logistics format so that the cost of supporting the product in the long term can be reduced.

Determination of requirements

Most important in determining product support are the quantitative requirements set forth by reliability and maintainability. These values are translated into qualitative requirements via the logistics support analysis, which supplements the maintenance concept in identifying tasks by echelon, repair policies, individual spare or repair parts and replacement frequency. These factors are combined to indicate supply support requirements for each geographical location and for the entire system. Major considerations include the following:

- Spare and repair parts covering actual item replacements are a result of corrective and preventive maintenance. This replacement includes both repairable and non-repairable items.
- Stock levels of spares compensate for repairable items undergoing maintenance. It becomes readily apparent that the support equipment capability, personnel and facilities directly affect the maintenance turnaround times and the quantity of additional spare items needed.
- Added stock levels of spare and repair parts compensate for the pipeline and procurement lead time for item acquisition.
- An additional stock level of spares compensates for condemning or scrapping repairable items. Repairable items returned to the intermediate or depot level are sometimes condemned because, through inspection, it is decided that it is not economically feasible to repair the item. Condemnation will vary depending on equipment utilisation, handling, environment and organisational capability. An increase in the condemnation rate will generally result in an increase in spare parts requirements.

Addressing spare and repair parts requirements from an optimum standpoint consists of solving three basic problems: (1) determination of the range or variety of spares, (2) determination of the optimum quantity for each line item and (3) evaluation of the impact of item selection and quantities on the effectiveness of the product. Items must be justified by establishing a demand prediction and identifying the consequences of not leaving the spare or repair part in stock. Some items are considered more critical to mission success than others. The criticality of an item is generally based on its function and not necessarily on its acquisition cost. However, the justification for these critical items may vary somewhat depending on the product and its mission.

The objectives in determining spare and repair parts requirements are to identify replacements; to determine replacement and repair frequencies, repair and resupply cycle times, condemnation factors and unit cost; and to develop supply support that will not impair the system by being non-operational due to supply and will minimise costs by eliminating unnecessary inventory and outstanding backorders.

During the early design and development stages, support objectives will be analysed through the use of models. The supply support capability will be simulated to arrive at an optimum balance between a stock-out reduction and the proper level of inventory. Stock-out promotes the cannibalisation of parts from other equipment (which may further deteriorate the system) and/or the initiation of high priority orders from the supplier. Both options are costly.

The purpose of the analysis and simulation is to arrive at the best approach. The results of a recommended solution are used in the initial provisioning of spares, repair parts and consumables. This analysis and simulation complements the study that the reliability engineer does through his predictions and failure modes and critical effects analysis. The combination of the analysis simulation and prediction techniques assists in establishing the required levels of spare parts needed to support the specific product in question. In other words, the appropriate materials are acquired to support the product for a specified time when the system first becomes operational. Re-provisioning is then accomplished as appropriate, using experienced data from the field to cover successive periods throughout the system's life. It is essentially the logistics engineer's responsibility to track the failures occurring in the field so that design optimisation in future equipment can be achieved. He is also responsible for ensuring that if failures are occurring because of items outside random failure bases, the trend analysis is established, and corrective action can take place based upon the results of whatever forensic engineering design studies are performed.

Provisioning and acquisition of material

Given the basic requirements for spares, repair parts and consumables, a plan is developed as part of the formal Logistics Support Plan for the provisioning of the appropriate material. Provisioning consists of the source coding of items, the preparation of stock lists and procurement documentation and the acquisition and delivery of material.* Usually it is not feasible to provision enough support for the entire life cycle of the product as too much capital would be tied up in inventory. The cost of inventory maintenance is high and much waste could occur, particularly if equipment changes are implemented and certain components become obsolete. In addition, initial provisioning is generally based on the estimated maintenance factors provided in the LSA (i.e. the replacement rates, repair and

* Source coding basically applies to the determination of whether an item is repairable or non-repairable and, if repairable, where it is to be repaired. This information is determined by the logistics support analysis, and the results are coded and included in the provisioning documentation. Procurement documentation includes identification of suppliers, pricing information, scheduling data, lead times, etc.

recycle times, pipeline times, etc.). These are only estimates derived from predictions, and as such may be in error.† Estimation errors will, of course, have a significant impact on the quantity of items in inventory; thus, provisioning should be accomplished at shorter intervals to allow for the necessary adjustments based on actual field experience.

But before this could actually occur, an initial supply of spare parts will be required to support the product that will be fielded before actual field experience data can be obtained. Therefore, as stated earlier, the logistics engineer should work closely with the reliability engineer and, using the results of the LSA, establish initial supply support requirements. These requirements can be integrated into the production cycle so that the procurement of parts can be done during the actual manufacture of production items, thereby reducing the total cost to the consumer.

Fig. 7.5 illustrates the possible differences between an early estimate of the maintenance replacement factor, the actual operational experience and a normalised set of values based on the operational experience. Fig. 7.5 shows the variation of estimates over time. An original estimate and its anticipated variation with time is made and used in the LSA and associated simulation efforts involving supply support. Initial provisioning is accomplished using this estimate as the basis for determining the quantity

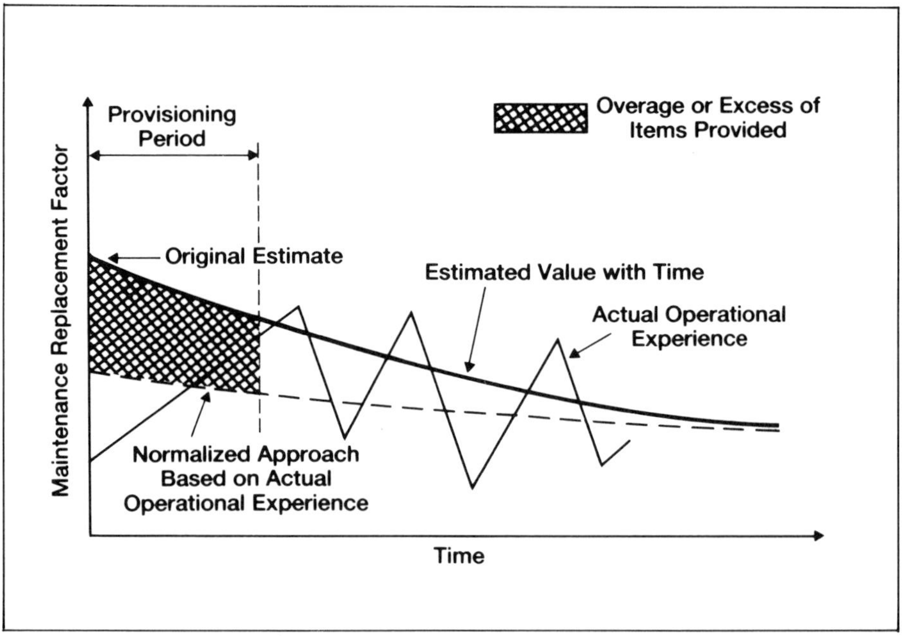

Fig. 7.5 The variation of maintenance replacement factor estimates over time.

† Experience on many systems has indicated that the correlation between predicted maintenance factors and actual field results has been rather low. It is hoped that as more data and experience relative to the input requirements and their relationships are acquired, predictions will continue to improve in the future, as they have in the past decade.

of different items in the inventory. The product is then deployed, and operational field data may indicate a different level of item usage. Actual usage may be erratic at first, but ultimately may be represented by a normalised curve as shown in Fig. 7.5. Assuming that initial provisioning is based on the period of time indicated, there will be an excess of items in the inventory. Maintenance factors can be adjusted for the next provisioning period, and the excess stockage can be used during that time. On the other hand, the provisioning of too little support results in the probability of causing the product to be inoperative due to stock-out, which can also be costly.

Generally, a realistic period of time is specified (e.g. six months, one year, two years, etc.) and enough spare/repair parts are procured to cover the demand requirements anticipated. Sometimes the quantities of different items procured will be adjusted for economic reasons. For instance, it may be feasible to acquire a set quantity of an item in order to realise a price break or to avoid excessive supplier production start-up and shut-down costs. If the quantity is excessive and the applicable items are not consumed during the selected support period, the residue may be carried into the next period of support.

The provisioning concept will vary somewhat between major high value items and smaller repair parts.* In addition, the procurement policies may differ for items with various usage rates. Fast-moving items may be procured locally near the point of usage such as the intermediate maintenance shop, while slower moving items stocked at the depot may be acquired from a remotely located supplier, as the pipeline and procurement lead times are not as critical.

The planning for major high value items is detailed and generally receives management scrutiny at every step in the process. This is required for both strategic and economic reasons. First, the supply status of these items has a significant influence on the supply levels of a large portion of the smaller repair parts in the inventory. Second, the value of the larger items is significant and may even exceed the total value of the hundreds of repair parts and accessories needed for their continued support. In other words, a relatively small number of items may represent a large percentage of the total inventory value. Thus, greater emphasis is placed on the computation of replacement factors and consumption rates to arrive at a true demand. In addition, the provisioning time periods may vary from item to item to ensure an economic order quantity that is compatible with the actual usage rate. In some instances, a given quantity of items are maintained in stock to compensate for repair and recycle times, pipeline and procurement lead times, etc., and new quantities are ordered on a one-for-one basis as existing items are withdrawn from the inventory.

---

* High value items are components with a relatively high unit acquisition cost, and should be provisioned on an individual basis. In addition, special packing and handling may be required. The classification of high value items will vary with the programme, and may be established at certain value (i.e. all components whose unit cost exceeds X are considered to be high value items).

Support and test equipment

Support elements are the tangible items of support that maintain products in the marketplace at a constant state of availability equal to the operational requirement. Support equipment is essential to the complete support of the product, and therefore must be planned and developed at the same time as the product. The development of the support equipment, and thus its influence on design decisions, is part of the overall objective of logistics support.

The definition of support and test equipment is all-inclusive: the tools, special monitoring, diagnostic and check-out equipment, metrology and calibration equipment, maintenance stands and servicing and handling equipment required to support scheduled and unscheduled maintenance of the end product. This definition is all-encompassing. It covers everything not necessarily associated with a specific prime element of product being developed to resolve a problem or mission requirement.

Support equipment is nearly as diverse as the primary product. It ranges from hand tools and simple gauges to highly sophisticated automated test and diagnostic equipment. Since it is quite possible that the support equipment may have to be conceived, designed, developed and produced concurrently with the product, it should be established early enough in the programme to be available when needed. Development of support equipment should not be relegated to a secondary position. These items are as essential as the prime element being developed. Without them the operation of the prime item cannot be verified and validated. It is essential that the tools and test equipment, particularly the latter, be handled in the same manner as the prime item being developed. The requirements to support it, notwithstanding all of the logistics support, should be in consonance with the needs for the prime item, because a failure in a piece of support equipment (especially unique test equipment) can have the same impact as a failure in the prime item development. It could cause the complete failure of the equipment in question.

Needless to say, support equipment should keep pace with the state-of-the-art technology. Maximum concern must be given to the hardware design so that needless functional duplication in the support equipment and procurement will be avoided. Therefore, prior to the start of support equipment development, every effort should be made to research the existing inventory for comparable hardware. Duplication can only increase the existing inventory and result in increased support requirements from which no real benefits can be derived.

A secondary consideration should be simplicity of design. Complex or difficult equipment would tend to pyramid into unneeded additions, and this should be avoided.

The importance of early life cycle planning of support equipment cannot be over-emphasised. Early consideration will influence design, built-in test and diagnostics, and trade-offs that could benefit availability and maintainability.

Specific approaches to support equipment requirements

As a result of any parametric design studies and the identification of specific technical approaches, refinement can initially be applied to the support equipment requirements for each design. The objectives of this refinement will be more defined support equipment requirements for each specific technical approach. For practical planning and support, each requirement falls into one of the following categories:

- Existing support equipment that can be used for some requirements.
- Existing support equipment that can be modified.
- Support equipment that must be developed.
- Support equipment that will be developed as peculiar to the end item.
- Built-in test and diagnostic support equipment in the end item.
- Automated or semi-automated support equipment.

Support equipment trade-offs

Trade-offs between required performance and the establishment of support equipment requirements may be desirable. This is, of course, based on the premise that such trade-offs contribute to the product's needs and cost effectiveness. Designs developed to optimise reliability must be compatible with the principle of good maintainability. Trade-offs to optimise support equipment requirements and resources should not degrade end item performance and/or reliability. A firm requirement for reliability or maintainability may require that support equipment requirements be subordinate, thus leading to the possible reduction of availability.

Revisions to support equipment requirements

The support equipment requirements should be reviewed periodically so that they do not become more complex than the end items. Revisions to previous recommendations should incorporate the results of advanced development activities, trade-offs, technical design changes, support equipment development requirements, advantages in new selection of existing support equipment and anything that will enhance cost effectiveness.

The revised requirements include any new plans to ensure that:

- Planned support equipment will be adequate within the tolerance required to ensure continued reliability and maintainability of repaired or overhauled products.
- Support equipment operation will not require personnel skills beyond those that can be practically made available.

- Calibration requirements, implications and implementations will have been considered in relation to the end item and support equipment.
- Development requirements for new support equipment will have been explained and justified.
- Proposed support equipment will include only realistic, effective requirements and exclude items insignificant to practical support.

Special test equipment

Special test equipment cannot be developed until the prime item is reasonably well into the FSED phase. What does this really mean? In essence, the logistician must work closely with the design engineer to ascertain whether special tools or unique test equipment will be required to support the product once it is deployed. Normally, the production of special purpose test equipment is frowned upon because this tends to drive up the cost, to make the prime item equipment unique in nature and make it more difficult for the end user to apply standard tools and test equipment in troubleshooting the end item.

In the past, it has been normal to delay the development of tools and test equipment until the prime item design has been frozen. This causes the test engineer or the logistician to take a hard look at what is needed to support the specific procurement in question. By waiting until the design has been frozen, the risks associated with the generation of test equipment will always be reduced. The problem with establishing special test equipment is, as was earlier mentioned, that the cost of the programme will begin to escalate. Special training programmes will be required to ensure that the maintenance personnel appreciate the application and use of the special equipment in question. The use of test programme sets is, perhaps, a lot more amenable to the services than attempting to build special or unique equipment to support the prime item.

The designer should carefully examine what is being put together and the amount of special equipment that will be required to support the design.

Minimising special support equipment becomes an extremely important aspect of the logistician's task. Early identification of the types of test equipment that should be used and the criteria that have been established to ensure that standardisation is maintained to the greatest possible extent is the logistician's prerequisite. Allowing the design engineer (though it may appear to be a sound choice at the time) to create special equipment may cause the logistician much more heartache than is worthwhile. Even though this appears to bring in more money to the manufacturer, in the long term it may cause him a lot more aggravation because it is an additional piece of equipment that he may be required to warrant. Again, the manufacturer's risk becomes high because he will have to support this equipment just as he has to support the prime item equipment.

Developing special test equipment also becomes something of a nightmare because the logistician should ensure that the test equipment is available when the prime item development hardware is distributed. The test equipment is essential to ensure that all of the required testing is inherent in the design, or that the equipment is made available at the appropriate level of maintenance in consonance with the delivery of the prime product. Again, this is a scheduling problem, and creates risks for the logistician and the design engineer. The risks from the design standpoint may be low, but the risk of not being able to deliver it on time could be exceptionally high. If it is not delivered along with the prime item, there may be a number of problems associated with supporting the prime item once it is fielded. It could result in the product being sent back to the manufacturer, causing him to accept under warranty the necessary and associated costs of repairing it.

Manufacturers sometimes develop test equipment so that they can actually test the prime item during the production phase. This does two things: (1) it ensures that the test equipment is operational and that it satisfies the requirements of the prime item and (2) it establishes the training techniques necessary to ensure proper hook-up and operation of the test equipment in conjunction with the prime item. But what it does not do is to minimise the total cost to the end user.

Table 7.2 shows the factors involved in the selection of test equipment. This table was prepared after a number of interviews, to ascertain what specific characteristics were critical to the various supporting functions.

It was not uncommon in the late 1950s and early 1960s for manufacturers of military equipment to produce as many different types of non-standard test equipment as they could use because this had a tendency of increasing the profit margin and causing a large proliferation of uncommon special test equipment that was being fielded by the Government. Because a limited amount of money is now becoming available for the development of prime item equipment, it has become increasingly difficult for manufacturers to try to sell the Government specific pieces of unique equipment to support the programme in question. Built-in test and diagnostic capabilities are beginning to replace special test equipment. If appropriately performed, the diagnostics will be able to troubleshoot the prime item down to the lowest replaceable assembly. This gives the maintenance technician the ability to remove and replace the faulty part or component at the lowest possible level without having to utilise any special test equipment to direct it down to that specific level. Special test equipment would then be used at the depot. As mentioned before, test equipment that was used by the contractor in developing the test profiles and test procedures will in essence be used again by the contractor to verify and validate that the equipment has been repaired to its operational state.

It must be noted now that the proliferation of specific test equipment and unique equipment to support a prime item is frowned upon by Government procuring activities. From the commercial point of view, any unique equipment that will be required to support a commercial product is also treated cautiously because it adds a level of complexity to the support that will be required to operationally test the product. Most consumers would

*Table 7.2 Factors in test equipment selection**

Factor	*Element*	*Rating*		
		Built-in	*Special purpose*	*General purpose*
Maintenance technician	Personnel acceptance	High	Medium	Low
	Personnel safety	High	High–medium	Medium–low
	Complexity of test-equipment operation	Low	Medium	High
	Time to complete tests	Least	Medium	Most
	Personnel-training time	Least	Medium	Most
	Tendency to over-depend on test equipment	High	High	Low
Physical factors	Limits on size of test equipment	Minimum limits; depends on prime-equipment and application		Maximum limits; limited by portability
	Limits on weight of test equipment	Minimum limits; depends on prime-equipment application		Maximum limits; limited by portability
	Complexity of 'wiring in' test equipment	High	High	Low
	Need for additional test point in prime equipment	None	None	Many
	Wanted space in work areas	Least	Some	Most
	Storage problems	None	None	Many
	Need for traffic considerations	Low	Medium	High

Table7.2 (continued)

Factor	*Element*	*Rating*		
		Built-in	*Special purpose*	*General purpose*
Maintainability and reliability	Probability of test-equipment damage	Low	Low	High
	Probability of damage to prime equipment caused by testing	Low	Low	High
	Effect on prime-equipment operation of repairing test-equipment failures	Some	Slight	None
Logistics	Cost to incorporate test equipment	High	Medium–high	None
	Test-equipment procurement time	High	Medium	Low
	Design-engineering effort	High–medium	High–medium	Low
	Compliance of test equipment to same specifications as prime equipment	Must	May	May
Application	Advantage of long duration and high-frequency usage in given location	High	High–medium	Low
	Versatility of application	Low	Low	High
	Opportunity for incorrect usage	Low	Low	High
	System adaptability to new test equipment	Low	Medium	High

* From US Army Material Command Pamphlet, AMCP 706-134, August 1967.

tend to shy away from anything that is a lot more complex than the consumer or his personnel are capable of handling. There is also this tendency when dealing with anything unique, which could cause the consumer to purchase a specific maintenance agreement and could result in a frequent need for field service personnel to maintain a piece of equipment that was advertised as being highly reliable.

This ultimately causes excessive downtime, additional training and more costs. Tools can be categorised in the same fashion as any special test equipment necessary to support the end item. Any special tool that is required to support the end item create additional problems for the maintenance engineer if there is a possibility of their being misplaced or lost. Standardisation in this category is of prime importance. Logisticians must make sure that there is nothing unique from the standpoint of the type of screwdrivers, phillips heads, or allen wrenches or any of the necessary support tools that the technician will need to ensure that he can take the end item apart and put it back together. This also applies in removing or replacing failed parts in the test equipment just discussed.

Training and training devices

This section will discuss training concepts necessary to ensure that the appropriate people are properly trained and transferred to the areas where the product will eventually be distributed. The operator, maintenance and supervisory personnel require training to ensure that all of the information required to support the product properly has been instilled into these people and that in-service organisational responsibilities (specifically, the organisational, intermediate and depot support of the activity in question, or the contractor support and interservice training requirements) have been appropriately identified. From a commercial point of view, this training also covers the field service personnel who will be responsible for supporting the product once it has reached the consumer (consumer being either a manufacturer who will re-use the equipment to support his facility, or a consumer responsible for handling the device in the manner for which it was designed). Commercially, technical training support also applies to the distributor who will be responsible for handling the returned product if and when the product fails. Field service personnel are those who will be responsible for supporting the product being used in other aspects of the manufacturing or support base; items like copying machines, blueprint machines, computers, etc., are the responsibility of a field service person.

Training is the responsibility of logistics management. It is incumbent upon the organisation manufacturing the product to ensure that appropriate training has been or will be given to the people responsible for support. This will cover different degrees and applications of issues associated with the specific product in question. If the unit has been built before and is just being upgraded, a minimum amount of training will be required for the people responsible for supporting it once it has been distributed to the end users. For all new design work, training becomes an extremely important facet to ensure that all of the personnel involved in support have the knowledge necessary to perform the requisite corrective action to bring the product back into operation.

What the logistics element manager for training needs to recognise early in the programme are the skill levels possessed by the people who are already employed by the corporation, or who will be using the equipment in the field. Matching the education requirements with the skill levels needed to operate the products is fundamental. If one tries to train an individual who does not have the necessary prerequisite knowledge, the problem will not be in the training techniques, but in the knowledge base of the person being trained. Prerequisites should be recognised early, and appropriate steps should be taken to upgrade the personnel skills so that the people available for future maintenance will have the necessary skills available when training on a new device commences.

The question is, how to establish the prerequisites of the individuals who will be responsible for supporting the fielded equipment? If it is within a corporation's purview, it should be able to identify the skill levels of the people already on hand so that the necessary additional skills can be quickly brought to bear as required. If new personnel are required to support the additional efforts, the interview process will need to ascertain their capabilities.

If this product will eventually be used in the commercial or military environment, who will be using it will be of the utmost importance to the manufacturer. For example, if the equipment will be deployed in a military environment and the knowledge base of the personnel who will operate or maintain it is not known, the depth and breadth of training will be very difficult to establish. The primary requirement from the manufacturer's point of view is to attempt to determine the skill levels of the personnel (as discussed on pp. 113-118). Once such skill levels have been defined and a requisite knowledge base has been established, preparation for training can commence.

If the product is to be sold within the private sector, the manufacturer will need to establish, through surveys of companies or people that may be utilising the types of equipment already manufactured, the levels of experience available. If the surveys indicate that inexperienced personnel will be required to support the hardware, either a minimal amount of training will be performed by the company's field service personnel, or in-house training will be provided at no cost to the users so that they will become familiar with the operations of the product in question. One of the most important elements here is to be able to make the training man/machine oriented. If the necessary approaches for maintaining the product are extremely difficult to perform, it is less likely that the end user will purchase the product without a maintenance contract. It is in the manufacturer's interest to make the repair of his product as simple as possible, for example with copying machines, blueprint machines and so on. Detailed information supplied by built-in tests is most appropriate for this type of task. Simple removals or replacements, resetting of safety switches, etc., should be built into the product.

Once the personnel capabilities have been defined, whether in the commercial or military environment, the logistician should work very closely with the training personnel to ensure that diagnostics are easily readable, material is readily available and the appropriate test points are quickly accessible to the people maintaining the equipment. This becomes a prerequisite to the design alternatives. As mentioned in earlier chapters, the logistics manager must readily understand and know with whom he will be dealing when the product is distributed. If this is not done in the design phase, the questionable ability of an outside individual to correct a problem promptly (or to be trained to correct it) quickly becomes an untenable situation.

To support the training techniques and materials being prepared by the training organisation, appropriate training devices may also be required. Specific items that need to be addressed along with these are the application and use of the product. Hands-on training is extremely important; the sooner the individual who will be operating or maintaining the product sees that product, the better off the manufacturer will be. On-the-job training is sometimes acceptable and workable. However, depending upon the complexity of the product, actual training performed by the manufacturer may be much more beneficial. The end product thereby becomes one of the training devices to be used by instructors during the training process. The instructor should also identify the necessary tools and test equipment required to support the maintenance of the product. This equipment is essential so that, again, the personnel being trained will become familiar with the applications and uses of the parts that the manufacturer recommends be used to support the product.

Two specific levels of training should be of concern to the manufacturer. The first level involves the operators, and the second the maintenance personnel. The first course is the operation of the product being distributed. This normally occurs prior to the maintenance course, but could include both the operators and the maintenance personnel because they need as much insight as to how the product is being operated as the operator. The maintenance personnel essentially will run the diagnostic routines to identify which part within the equipment has failed or appears to have failed. The operator will utilise the equipment to generate and identify the results of what the product is supposed to produce. The operator should also be familiar with all of the knobs and tuning devices, if any, required to support the operation of the product.

From a military standpoint, the operator is the person who actually uses the product in its designed environment. Commercially, the operator is the end user, such as those who purchase televisions, radios, stereophonic equipment and other electronic appliances. We recognise that, from a commercial point of view, equipment operation must be very simple. If it becomes too complicated, the chances of having other people buy the product decrease.

For example: if the description of a switch or knob is not readily available, or if the definition supplied in the user's manual is not easily understandable, the level of discomfort and annoyance to the user becomes very high. He or she will become disgruntled and eventually tell friends that the device is too complex. As few knobs, switches or readouts as possible should be applied to the mechanical or electromechanical devices being manufactured. The less complex, the better off it is; the more complex, the harder it becomes for the consumer to understand and fully operate. From the military point of view, because of the complexity of the devices now being designed, the engineer, logistician and training personnel must ensure that the interface between man and machine is kept as simple as possible. The readouts and adjustments necessary to produce the required signals to support the mission of the equipment must be readily accessible. The readouts and adjustments should be quickly understood by the operator so that he has an appreciation of how to operate the product for which he will be responsible. The logistician, engineer and trainer should design equipment so that it does not require an engineer with a Ph.D to fully understand its operation and use.

From the maintenance point of view, training will be required at the various levels of support, whether organisational, intermediate or depot. Again, the maintenance technician will be required to have an appreciation of how the product is supposed to operate: therefore, he will need a level of operator training similar to that given to the operator. For if diagnostic routines have been built into the product, the technician should be able to readily appreciate and review the signals or readouts presented to him. Training the maintenance technician on the actual hardware with the test equipment that eventually will be distributed with the equipment will make his job that much simpler when he actually gets in a position of support.

The degree of training will depend upon the personnel and the complexity of the product. It is normally accepted that there will be a degree of hands-on training associated with the training programme. The amount of training will depend upon the complexity of the hardware. Normally, it should be a 60/40 division: 60% covering the theoretical aspects of the product, and 40% being the actual hands-on training techniques used to reinforce the information that has been covered in the theoretical training. Training on the test equipment is no less important than training on the prime equipment. It is important that training across the board be presented to the technician so that he is in a position not only to evaluate and determine whether there is a problem with the prime equipment, but also to determine whether there is a problem in the test equipment being used to ensure the proper operation of the prime equipment.

How does one determine how much training is required? That is a very difficult question to assess. It is also based upon the experience of the training personnel who will prepare the courses necessary to support the training activities. From a military standpoint, training is normally dictated according to the requirements document submitted as part of the full-scale engineering development or production phases. The government specifically states the type, depth and breadth of training programmes that it anticipates will be required to support the equipment. It is normally defined as the number of hours required to train the personnel, covering both theoretical and operational approaches. The contract documentation will also specify the type of support material or training devices that will be required to ensure that once the people being trained are deployed, they will have the appropriate material on hand to reinforce the training they received. This leads to the deliverable products associated with a training programme.

From a commercial point of view, depending upon the size of the product being manufactured and distributed, the manufacturer, as part of the total procurement, should establish training courses to be used by the customers (or by their personnel) to learn the use, application and unique attributes of the product. This ensures some control over the speed at which the new user will learn the materials. It also ensures that the manufacturer will have an appreciation of the amount of knowledge possessed by the personnel involved in supporting the equipment once it has been employed by the consumer. The manufacturer will also have the ability to identify what field service personnel will need to know to augment the training programme or to actually maintain the equipment, so that when they get into the field they will have some insight as to the depth of training that was given to the users and will have an appreciation of the types of problems that they may be running into.

It is not too farfetched to suggest that the people who are actually doing the field service work could be the training personnel as well. If this technique is employed, it gives them the ability to quickly learn the techniques associated with maintaining and troubleshooting the equipment. The dual role becomes a very interesting aspect in the logistics capabilities of an organisation. If employed, the programme will constantly be fine-tuned, and the application and use of training techniques and suggested problem-solving techniques will be integrated into the total training package.

This approach should also be considered by the manufacturers of military equipment, because the personnel who are actually in the field supporting the operation and maintenance of the equipment can readily identify the unique techniques necessary to ensure proper operation and maintenance of the prime equipment and make them available to the logistics element manager responsible for training. He can also do the same thing for the test equipment, if unique to the manufacturer. By closing the loop, this ensures the fine tuning of all training necessary so that, as additional personnel are processed through the manufacturer's facility, the best training programme can be presented. This is, in fact, the final step in ensuring that the product the manufacturer is producing will have the highest quality of operation and maintenance use possible. Many manufacturers' downfall is that they do not provide sufficient training to the operators and maintainers, who are the people who will establish the credibility of the application of the prime equipment once it is deployed. If the training becomes inadequate or if the approach is inadequate, once the equipment is fielded, dissatisfied operators and maintainers will determine whether or not there will be future purchases. The manufacturer should ensure that the training programmes are prepared and conducted as well as possible and to the level necessary to ensure that the personnel in question possess adequate operation and maintenance capabilities.

Technical documentation associated with a training programme

To be able to support a proper training programme, technical documentation should be prepared by the personnel responsible for conducting the training courses. Training material early in a programme is in the form of a technical training plan. This establishes what the training group is going to do, how they will conduct the training programme and what will be required to support it. It gives guidance, establishes programming capabilities and the necessary schedules to ensure that when the training course is conducted, and all of the appropriate documentation is available to support it.

It is suggested that some, if not all, of the following documents be available for both the instructor and the student at the time the course is presented:

- training plan;
- course outline;
- instructor's guide;
- technical manuals;
- other presentation material.

In conclusion, training consists of the processes, procedures techniques, training devices and equipment used to train civilian and military personnel in the operation and support of a new product. This includes individual and crew training; new equipment training; initial, formal and on-the-job training; and logistics support training.

Technical data

Technical data are defined as all recorded information, regardless of form or character, such as manuals, drawings, engineering studies or analyses, or even engineering notes that support the development of the specific design. As an example, computer programs are not considered technical data; however, the documentation of the computer program and its related software are. Also excluded from the technical database is financial data and any other information related to contract administration. Therefore, let us define technical data as the recorded information used to define a design and to produce, support, maintain or operate the specific product. Technical data ensure optimal product effectiveness and economy. Accurate determination of the requirements for the acquisition and timely utilisation of adequate technical data is as follows:

- Planning data requirements concurrently with planning systems details and services.
- Procuring data only when needed and when requirements can be economically justified.
- Providing data requirements for objective review by those outside the organisation.
- Promoting effective use of data with other ILS functions, i.e. configuration management, provisioning, reliability, etc.
- Promoting optimum uniformity and component data requirements, thereby avoiding the unnecessary cost of presentation and administration in facilitating the exchange of data.

This model is all-encompassing, but attempts to illustrate the need for communications and overall control of data within a specific organisation. The real significance of technical data provided with the product turns out to be the information necessary to generate the technical manuals. The data is used by the technicians, enabling them to remove or replace, ship or package the products that they receive or that they need for their jobs. Fig. 7.6 depicts the necessary information in the flow of technical data needed to ensure that a product and its related services are being produced in accordance with the requirements of the contract. At the peak of the pyramid is the programme management office, which is responsible for ensuring that flow of information between each of the functions responsible for producing or supporting specific aspects of the product is appropriate or adequate.

The downward flow of technical data consists of all documentation received by the contractor and the guidance prepared by the programme office that is necessary to ensure that each of the functions operates in accordance with the established requirements. This also includes the task statements and schedules, along with the budgets, that are needed so that each of the operating functions succeeds. For discussion purposes, let us assume that all of the lines that connect these functions are buses and that the information flows along the buses so that each of the functions that support the programme interfaces with the other functions to facilitate the accomplishment of assigned tasks. The flow of technical data proceeds upwards as well. It is imperative that data are generated by the individual functions flow up

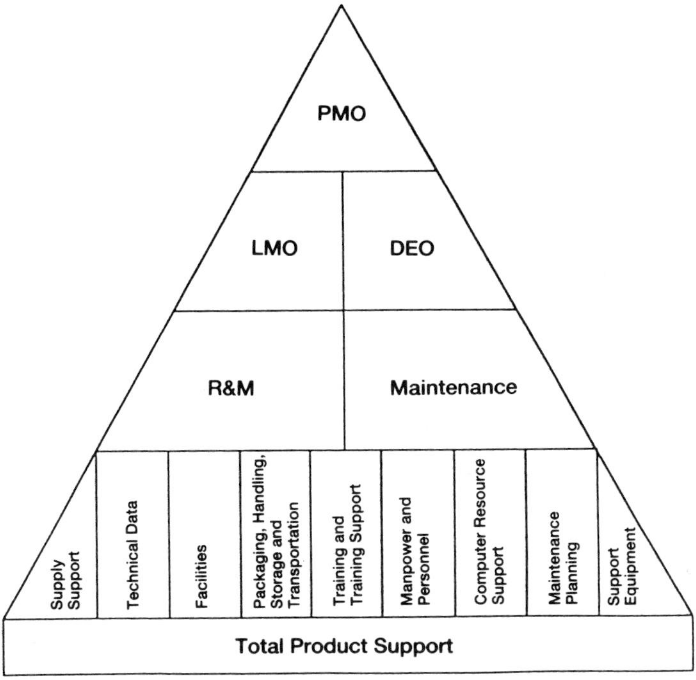

Fig. 7.6 Flow of technical data required. LMO, logistics management office; PMO, programme management office; R&M, reliability and maintainability; DEO, design engineering office.

to the management level so that each manager will have a full appreciation of what is being accomplished and whether it is being successfully done in accordance with the requirements established by the programme office.

The programme office normally uses the contract as the vehicle to identify what technical data will be required to be submitted to the procuring activity. It will also identify internal technical documentation needed to support the programme in assisting and making the appropriate decisions. This decision-making process is also the responsibility of the logistics managers as well as the design engineering manager. They both require technical data to support the proposed designs and to establish the basic configuration from which the final design is formulated.

Special requirements are defined by the programme office. The logistics manager's responsibility will be to initiate or review the technical documentation requiring the logistics elements managers to determine how best to schedule each of the requirements to meet the contract. The logistician or the logistics manager is responsible for tracking these requirements in accordance with the contract schedule to ensure that the data are prepared to meet the contract. One of the specific duties of the logistics manager is to establish a technical data group, which is responsible for ensuring that all of the technical data called out as part of the logistics requirements programme is prepared in accordance with the appropriate documentation.

Two types of data are needed for customer versatility. The first is the technical data that cover operating and maintenance support and are used during the design development phase. These documents support the equipment which flows from the procuring activity to their users' sites. The purpose of the technical data programme is to provide for the timely and coordinated development and distribution of the technical data necessary to conduct field operations. The planning for this data must be based upon information from equipment operations and maintenance planners that takes into consideration the design, support trade-offs, test demonstrations, production, operation and maintenance. The data just described are normally generated through the use of logistics support analysis record sheets. They are used to help support the data associated with the specific programme needed to ensure that the logistics material is properly documented. The base of the foundation in Fig. 7.6 is total systems support. In total systems support, the flow of data essentially involves all of the documentation associated with the individual logistics items that provide the link between product operation and performance and the personnel who are responsible for operating and maintaining it.

The second type is directly related to the management data and are either delivered to the contracting office or maintained by the programme office to help establish the audit trails necessary to support the product. These data are provided to identify the aims of a programme during its development cycle for management so that timely, precise responses can be made. In other words, the data flows from the base up to the top of the pyramid, and is normally reviewed by each of the functions above the foundation of the pyramid. These data are used to help assess the areas in which possible problems may arise, and also used to help solve those problems early enough so that appropriate decisions can be made. The flow upward and downward within the pyramid is extremely important. This flow maintains the continuity between the line management and the staff within the organisation and shows that the information is properly transferred so that the appropriate decisions can be translated into actual design.

Determination of intended use of data

Specific requirements for data should be established as early as possible during the concept formulation or contract definition phase. Information available from prior planning should be used in decision making to determine follow-up or logistics data requirements. Data requirements will be determined on the basis of the intended use of the data, with careful consideration given to the immediate and possible future use of the products, materials and services to which the data relates. It is most important to determine the programme type so that consideration can be given to the support data required to maintain operation during both acquisition and deployment.

Checklists should be made to assist in realistically determining the data that the programme requires for completion. This is done to ensure that the final end user or consumer has the data required to support his needs. By establishing realistic data requirements early, he may save later expense and possible difficulties involved with the need to procure or generate additional technical data.

Technical data development

Fig. 7.7 will assist the reader in understanding the flow of technical data that normally occurs during the development of a product. First and foremost is the engineering and technical data generated to support the logistics profile. These are used to highlight some of the specific requirements needed to ensure that the support aspects of the programme have been covered and that they identify some of the required verification and validation (V&V) steps to be performed so that the technical data satisfies product use. Thus a closed loop has been established between the actual formation of the documentation and the resolution of problems found during the V&V phases. It also includes the assurance that all the technical data are current and satisfies the configuration baseline established for the product that is being fielded.

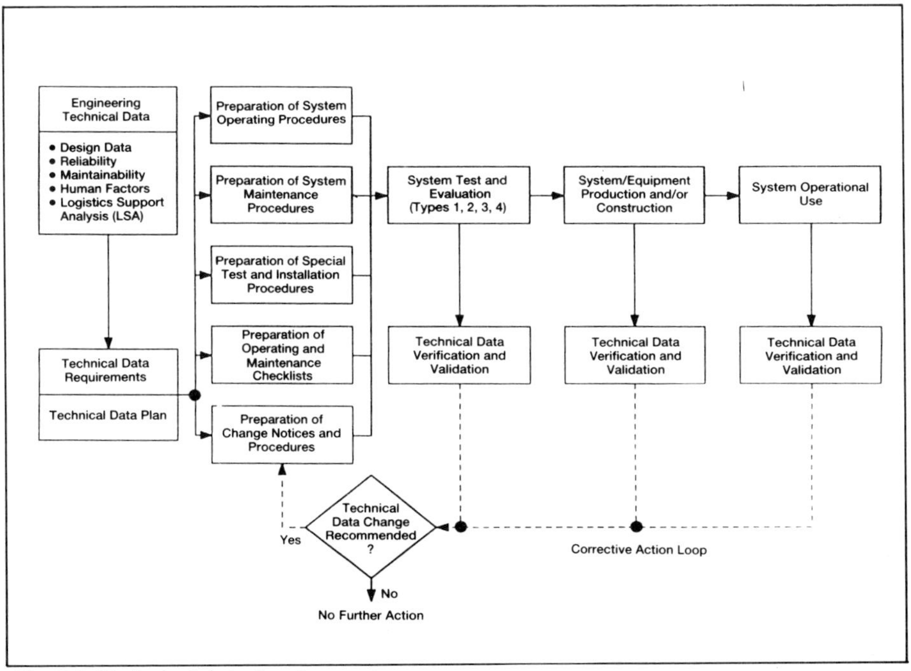

Fig. 7.7 Technical data development.

Equipment publications

The rapid advance in technology and the technical complexity of modern products have increased consumer concern. The amount of information to be conveyed to both the user and the maintenance technician has also escalated, whilst there has been no increase in time available to train the user in its application. This task of communicating the continually increasing volume of technical information to relatively non-technical operators and maintenance personnel is the responsibility of equipment publications. The success of ILS is directly affected by the proper interface between design and publications. In the USA the tendency is to say that when the paperwork equals the weight of the product you are designing, the product has been completed. In other words, for the development of aircraft, like the B-1, the weight of all of the technical publications and of the technical data have to equal the weight of the aircraft before assuming that the product has successfully met its requirements. This concept is a little far fetched, but in truth, because of the requirements being dictated by the customer, it is becoming very difficult for the manufacturer to be totally responsive to the needs of the programme without producing this amount of paper. This must be curtailed, and it is up to the logistician to sit down with his counterpart (either the prime contractor or the procuring activity) to coordinate efforts to ensure that only the absolute minimum needs are maintained and achieved through required documentation.

If a device has been built before and off-the-shelf material is available using a commercially acceptable approach, then this material should be suggested for use by the consumer. It is not necessary to re-do something that has already been done, and the contractor can illustrate this by allowing the consumer (be it the Government or the final applications user) to review the documentation prior to contract award, or at least prior to defining the data package, so that he will understand the project. By doing so, minor modifications to the existing documentation, if required, can be made quickly at a very low cost. This will also eliminate some of the bulk of material that will be required to support the product. Because of the size of the quarters in which some products will be installed, it may become too cramped to store all of the technical documentation necessary to support it. Therefore, it is up to the procuring activity to incorporate into the end item automatic features of all types that are economically and technically sound to reduce the volume of printed instructions. Further efficiencies to convey information to the end item user can be achieved through the use of advanced technology, such as micro-cards, laser disks and other microelectronic devices capable of storing large volumes of technical data.

One must never forget that the equipment manuals and publications produced by the manufacturer are his point of contact with the person who will be responsible for operating and maintaining the equipment once it leaves the facility. This documentation will determine the manufacturer's reputation. If the quality of the product is insufficient, it will affect his ability to sell additional products in the same market. The accuracy, clarity, and specificity of his documentation are essential. Any mistakes or inaccuracies in the documentation could represent a major fault in the operation and use of the product. This could result in both harm to the end user and the inability of a maintainer or operator to successfully complete his work. It also results in many false alarms or failures. Therefore it is the contractor's responsibility to ensure that the technical data satisfactorily identifies and portrays the product that he is distributing, whether it be to the consumer or to a military agency. Either way, he is ultimately responsible for what the printed matter says. This responsibility falls on the logistics manager's shoulders, because he is responsible for validating and verifying the technical documentation. Remember, proper preparation prevents poor performance.

Computer resources support

Since the advent of computer resources, the logistician has been encumbered with another major responsibility: to ensure that all of the changes associated with computer programs are properly documented and maintained. It is also his responsibility that these changes, as they are needed and/or implemented, are done in an appropriate manner so that when a changeover does occur it does not have an effect on the actual product's operation. In earlier years, a change in the design would usually require only a change in a piece of electronic equipment. But nowadays, with the advent of the computer, a reassessment of requirements could mean a change in software, which may affect the entire design. A very simple software change could have an impact on absolutely everything that is being accomplished with the design itself. If the software change is not properly coordinated and implemented, it could result in many false alarms in the product that he distributed or in any interconnecting equipment. Surprisingly enough, it is not as easy as one thinks to document a specific change in software. Notes can be associated with the software and incorporated into the design package, or engineering notes can be assigned to a specific product that is being generated in support of the software.

Either way, it is the logistician's responsibility to ensure that the appropriate documentation is always available just in case there is a problem; otherwise, there is no alternative but to return to the original baseline.

It is extremely important, especially in this case, that an audit trail is always maintained – that the record database of the software programs is retained. The computer support resources that the logistician is required to review are those necessary to ensure that the computer software being generated will be properly identified, annotated and documented. When the equipment has been finalised he will be able to track what software was generated, where it is located and how well it is operating. The logistician's approach to computer resources is no different from his approach to the hardware, except that it becomes a little more dynamic in nature. Hardware is rapidly identified and the functional aspects of it can be quickly specified. Tracking it from one functional element to another is less dynamic than tracking software designs.

The techniques that must be applied to the software by the logistician become a little more complex than those for the hardware: techniques for analysing problems being experienced, identifying the types of problems that may result from software design versus hardware design or trying to establish the basis for the probable fault. The diagnostic routines for hardware may not be directly applicable to software. There must be specific diagnostic routines that are designed to help check all of the software applications being used in the product. Today's software technology drives hardware technology. Logisticians must carefully evaluate what that means and what impact it will have on the support system. Software comprises an ever-increasing portion of many products; therefore, one must consider how best to handle the support aspects and where and when to make the appropriate modifications.

The technologists are beginning to recognise that many software elements, specifically the algorithms, cannot be used over again. Software techniques are not necessarily standardised. The application is standardised but the actual code may not be able to be used from one application to the next. Major software expenditures continue to be experienced, therefore care must be taken to ensure that minimum levels of software are all that is required for the application of the product in its intended environment.

The logistician should attempt to establish a means by which he can identify the types of modifications. This may be required during the operational phase of the system. He must be able to ensure that the diagnostic programmes are fully evaluated, and that any deficiencies are corrected prior to future deployment. The operational phase of software is critical, for without a full appreciation of the software's capabilities and the types of modifications available to ensure its proper use, the logistician is going to have a major problem in determining how best to modify or correct any deficiencies that may be found during the product's operation.

Once the equipment is operating properly, it is up to the logistician to ensure that a software configuration baseline has been established. This provides a departure for future changes. It will ensure that as changes are implemented, the logistician has at least an ability to refer to an operational baseline that was working prior to the change. One of the most important features in software configuration support resources is the ability to perform testing to ensure that all software modifications will work in accordance with the specifications.

A standard procedure should be established and maintained to ensure that as modifications occur, the logistician and his support personnel can perform the necessary tests to validate and verify that the software will meet its requirements. If there are problems in the implementation of modifications, the logistician can refer to the original baseline and ensure that there will be no implied changes to the software until the new modification is operating in the desired environment. This is important because the logistician and his staff need to be able to identify and document any required change. Even though the field personnel identify changes, the logistician and the design engineer must be able to verify that these changes are needed and that they can be implemented with minimum impact to the operating system.

The second most important factor is being able to know where all of the operating products are and what the configuration baseline of each product is. In addition, he must recognise that the configurations differ among the products, even though they were designed to a standard base. Any modifications implemented by the maintenance and operational personnel by 'bootlegging' (loading the software with no backup software or documentation) them into the product could have a direct impact on any change that the logistician and design engineer wish to implement. Therefore, periodic audits of the equipment, depending on its size and complexity, should be performed to ensure that the status of the software has not changed and any change will be accepted by the product with minimum impact.

The logistician must ensure early in the programme that his supporting documentation adequately addresses the support planning for a user programme whether for firmware or software. This planning should highlight documentation training, support equipment and facilities required for the software. ILS planning should also be consistent with the systems engineering software and configuration management procedures to ensure that the software is implemented and maintained at the level of operation specified by the operating documentation.

All of this should be done in conjunction with a structured software development process. Even though there are changes to the programme, this process should be adhered to and maintained so that any of the modifications previously discussed can be properly implemented. Not enough can be said about the importance of being able to control computer software. Without such control, the software could become unrealistically expensive, and have a major impact on the reliability/maintainability of the product.

Packaging, handling, storage and transportation

Physical distribution and transportation

Physical distribution and transportation are not always integrated forms and functions in corporations. Together they can represent as high as 70% of the total logistics costs for a company, and transportation costs alone may account for as much as 40 to 50%. Managerial decisions for physical distribution and transportation, often referred to as 'PD&T', directly influence all functional areas of the business. For this reason, it has an important representation in the corporate strategy. As early as 1982, Temple, Barker and Sloane, Inc. wrote that with the ending of transportation regulation, a proactive management to PD&T is needed. Identifying and solving problems requires the 'use of strategic and operating plans, budgets, carrier negotiations and formal reviews, effective communications, and human resources' for competitive advantage.[1]

Physical distribution and transportation strategy

These imperatives require the ILS manager and the logistician, in conjunction with top corporate management, to develop and support an effective transportation strategy. Coyle, Bardi and Langley define this strategy as a component of the firm's logistics strategy pertaining to the purchase, monitor and control of freight transportation service.[2] Transportation strategy enhances the firm's overall strategy.

Fig. 7.8 is adapted from Temple, Barker and Sloane, Inc. as quoted in Coyle, Bardi and Langley.[3] The figure graphically demonstrates how the strategy for PD&T is incorporated into the firm's logistics strategy, and how both relate to the corporate strategy. Note how all three are affected by the external environment.

To execute this strategy management must be proactive, and well versed in analytical skills and problem solving. Although a rare trait, the logistician and his analysts should be micro-intuitive, in the sense that they can foresee the effects of alternative decisions. The authors know of no colleges or universities that teach a developmental course on micro-intuition. Useful, user-friendly information, planning and operating systems are required. Finally, a strong sense of affability and negotiating skills are necessary.

Physical distribution and transportation functions

With a well-formulated and articulated PD&T strategy, the PD&T manager's job is much easier, and begs the question, 'What are the functions of PD&T?'

Physical distribution and transportation are subfunctions of logistics whose complexities require highly qualified professional logisticians. Such complexities include:

---

[1] Temple, Barker, and Sloane, Inc. (1982) *Transportation Strategies for the Eighties* (National Council of Physical Distribution Management), p. 150.
[2] Coyle, Bardi and Langley (1988), p. 393.
[3] Temple, Barker, and Sloane, Inc. as quoted in Coyle, Bardi, and Langley (1988).

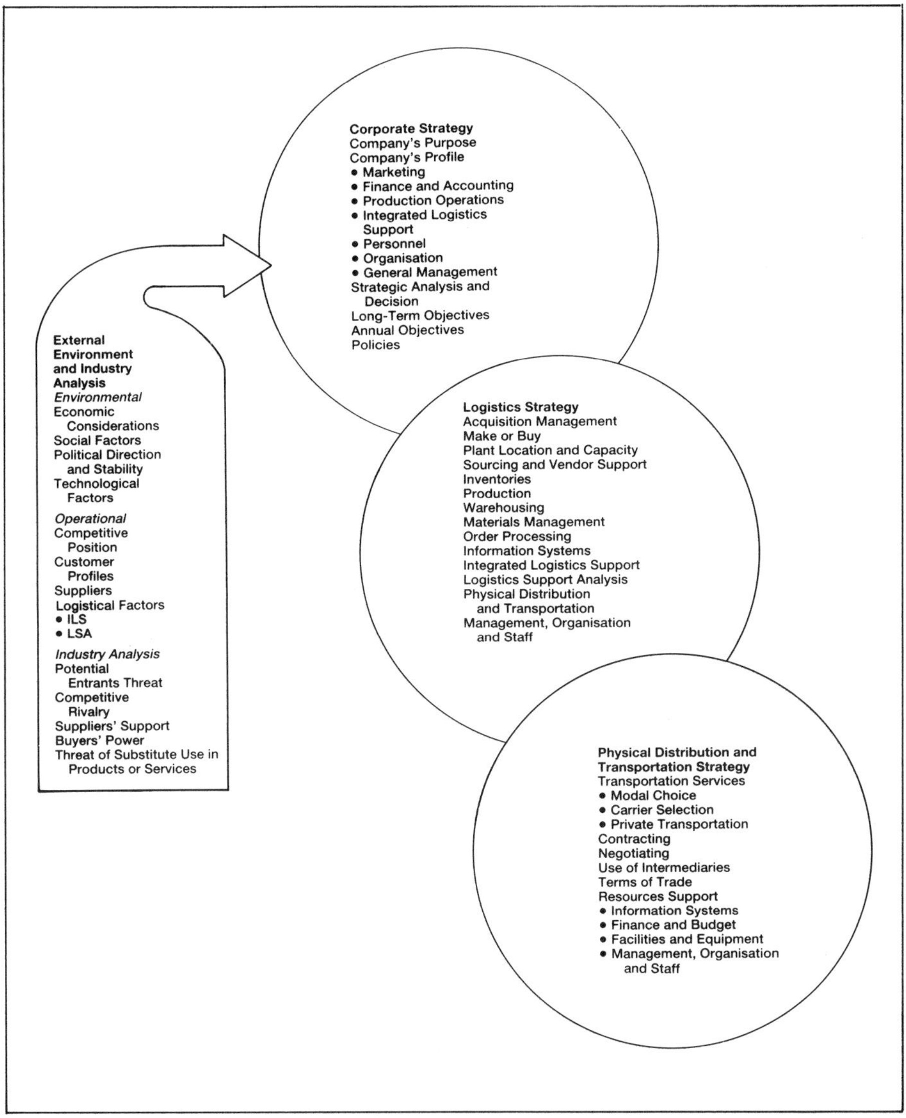

Fig. 7.8 Logistics corporate strategies.

- selecting optional modes;
- determining which carrier to choose from numerous vendors;
- meeting requirements within the regulatory structure;
- obtaining the best rates from a price structure that has no rationale; and
- balancing the firm's objectives for profitability and sustainable growth against the requirements for quality, excellence and customer service.

The importance of physical distribution and transportation becomes readily apparent when we examine the complexities outlined above.

Physical distribution and transportation costs are the largest component of any commercial firm's total outlay. For example, a study of 160 companies conducted by Herbert W. Davis in 1986 indicated that physical distribution costs represent 7.76% of each sales dollar, and transportation accounts for 3.14% of sales. No other factor in the production of finished goods represents such a high percentage of sales dollars.[4] These percentages certainly indicate the relative high importance of conducting physical distribution and transportation functions effectively. In macro terms, PD&T accounts for 7% of United States' GNP.

The nature of PD&T

Physical distribution and transportation consists of storage of raw materials and finished products, transportation of raw materials and assemblies needed for development of the product; materials handling; inventory control; warehousing; inbound, internal, and outbound flows of materials and service components; order processing and customer service; administration to include determining inventory carrying costs; transportation to wholesalers, retailers, and users; determining user satisfaction, waste disposal; and managing obsolescence in terms of materials, equipment, products, and services. PD&T is essential to the viability, marketability, cost effectiveness and excellence of product and service functions. It fills the gap between the producer and the user. To carry out PD&T well requires responsiveness, information systems, keen management skills and the employment and use of a total systems approach. PD&T creates time and place utility in goods and services.

Some theorists consider purchasing as dealing with inbound materials, and distribution as being concerned with the outbound flow of materials and services. This common misinterpretation also exists in industry, and requires some explanation as to why PD&T costs are as high as they are. Procurement specialists, including purchasing agents, are rarely professionally experienced enough to perform inbound distribution and transportation functions well. Furthermore, to subdivide inbound and outbound flows of materials and services by speciality or functions tends to ignore the effectiveness of managing the internal flow of materials, and results in even higher costs to the firm. Overall, it is estimated that in excess of 10% of the US total labour force is involved in physical distribution and transportation functions.

PD&T managers concern themselves with four basic issues.

1. *Finished good inventory positioning.* Positioning generally applies to where to stock the inventory of finished goods.

- Forward positioning applies when locating stock in warehouses, wholesalers or retailers close to customers.

---

[4] Davis, Herbert W. *Davis Database* 1986, Vol. II, No. 7.

- Rearward (sometimes referred to as backward) positioning applies to retaining or holding the stock on the manufacturer's premises. Most industries opt for forward positioning because it provides:
  - faster delivery times,
  - potential lower transportation costs, and
  - responsiveness that enhances the firm's competitive advantage through stimulated sales.

2. *Transportation mode selection.* Government deregulation of the transportation sector has increased competition, which resulted in vast technological changes in modes and compounded the PD&T manager's decision in which mode to select. There are five principal modes of transportation: rail, truck, water, air and pipeline. Combinations of these five provide for intermodalism.

Rail transportation is used to move large quantities. It is ideally suited for shipments of bulk and raw materials. Rail provides intra-country capabilities, providing rail lines are maintained.

Highway transportation is the most popular of the modes because of its versatility and responsiveness. Re-handling during pick-up and delivery is minimal. For short distances transit times and rates are often less than rail.

Water transportation has slow transit times, but provides the capability to move huge and heavy amounts of material at relatively low cost. Its biggest drawback lies in the inaccessibility of inland communities and manufacturers.

Air transportation is quick and flexible. However, it is the most expensive of all modes, and it is limited to lightweight shipments when compared to other modes. Its popularity is increasing dramatically, but still represents only 1% of all shipments.

Pipelines are low in operating costs. They have a largely undeveloped potential, and their use in the USA is generally limited to liquids, gases or solids in slurry form. Some innovative firms use pipelines to ship fish and coal.

Intermodal transportation refers to combinations of the five and to special service modes and hybrids, such as air express, bus service and piggyback. Increasingly, containers are used as intermodal transport. The flexibility of intermodal containers is somewhat limited because of the popular preference for lightweight aluminium containers that cannot be moved about fully loaded by forklift.

Another decisive element facing the PD&T manager is the type of carrier. Carriers may be private (owned and operated by the manufacturer); contract (hired from firms in the carrier business); or common (who serve all customers without discriminating).

3. *Scheduling, routing and selecting a carrier.* Daily control of freight involves many issues. Among them are development of the shipping schedule in coordination with purchasing and the operations departments;

trade-offs between lower transportation costs and customer service; and consistency with the integrated logistics support (ILS) framework. The huge variance in rate structures makes the PD&T manager's decisions complex.

4. *Inventory control.* Inventory is subdivided into three categories: raw materials, work-in-progress and finished goods. Inventory management is relatively simple for the small retailer, but for larger retailers and manufacturers inventory is costly and its management who deal with components and subassemblies find inventory control challenging and complex.

Several innovations have occurred to attempt to make inventory control more manageable, such as part commonality (where the same part may be used to manufacture several subassemblies or products), or *kanban* (a Japanese system that provides for the use of cards or computer record to control the flow of production and inventory).

Inventory relies on types of demand. Independent demand refers to items not relying on the demand of other items. It is affected by market conditions and generated by customer demand. Dependent demand relates to all the parts, components, subassemblies and assemblies necessary to develop a product. Independent demand is forecast, while dependent demand is based on production plans for the product or service.

Inventory may be measured in one of three ways:

1. *Aggregate inventory value* is based on the total costs. It is calculated by multiplying the total number of parts or components by the cost for each, and then totalling the entire categories of parts, components, etc. This figure demonstrates how much of the firm's assets are tied up in inventory. Retailers and wholesalers average around 75% of their total assets, while manufacturers average 25%.
2. *Weeks (or days) of supply* takes the aggregate inventory value and divides sales per week at cost into it, i.e. only the cost of finished goods sold. For example, assume an average $24 million in inventory this past year, and the cost of goods sold was $8 million. Measured inventory is:

$$\text{Weeks of Supply} = \frac{\text{4 million}}{\text{\$8 million/52 weeks}} = \text{26 weeks}$$

3. *Inventory turnover* is calculated by dividing annual sales at cost by the average inventory value on hand during the year. For example, using the above figures:

$$\text{Turnover} = \frac{\text{8 million}}{\text{4 million}} = \text{2 turnovers/year}$$

The optimal inventory level is difficult to determine, and is based on the individual firm's history of success. As a guide for comparative purposes, about five or six turnovers per year is the industry average, while high-tech industries average about three or so. Some writers report that Japanese auto industries indicate about 40 turnovers a year.

Inventory turnover is critical. Large inventories can provide better customer service; better facility and labour utility; and cheaper transportation costs. Smaller inventories improve cash flow position; lower storage costs; lower property tax; and lower insurance costs. These factors must be considered, as well as the varying types of inventory costs, such as holding (carrying costs); ordering costs; transportation costs; management's importance for providing customer service; and utilisation rates for labour, equipment, together with the facility.

When a firm employs *kanban*, the inventory level is determined somewhat differently. The amount of inventory for each part (i.e. inventory on-hand and scheduled-in parts) should be equal to the authorised number of parts in each container multiplied by the number of cards for that part.

Conclusion

High inventory levels are costly. The extreme alternative, poor customer service, costs market share. These are major considerations. The physical distribution and transportation manager who is knowledgeable of ILS can avoid making an inappropriate decision on choices and selections which are costly. His or her decisions on the positioning of finished goods inventories, transportation mode selection, shipping schedules, routes, and carriers and inventory control means the difference in the firm's ability to achieve excellence, quality and profit. The principal way this is accomplished is through the effective implementation of a PD&T strategy which is integrated into the firm's logistics strategy. These, in turn, are integral to the firm's corporate strategy.

Facilities

Surprisingly enough, the facilities function turn out to be the stepchild of the logistics functions. What they are really part of, and what they do, is only given lip service. Their maximum potential is not rated until somebody realises that a facility's requirements statement is absolutely necessary, or until such time as the equipment is deployed or distributed and it is discovered that the facilities that are already available are too small to handle the need. It is time that facilities take less of a back seat in logistics planning and that the facility's requirements and the planning necessary to support them become much more pronounced in the logistics circles than they have been in the past.

From the commercial standpoint, facilities planning plays much more of a dominant role than it has in the military environment. Facilities, from a commercial point of view, are the areas in which the equipment is distributed and is properly housed, the materials support is made available, the inventory and distribution systems are in place, the warehouses are properly supported and additional space has been provided to meet the needs of any future device. The commercial houses take care of facilities early in the logistics cycle because they are essential to the distribution of the product to the consumer.

For the military purposes, facilities encompass the real property assets required to support the product and the studies that define types of facilities or facility improvements, location, space needed, etc. Therefore, the objective of the ILS facility plan is to ensure that the required facilities are available to all of the organisations responsible for operating forces and supporting activities at the time they are needed. Because it takes so long to implement a change in a facility or to develop new ones, early planning must be done. In the USA a minimum of five years is normally required from initiation of the military Programme Objectives Memorandum (POM) process until a usable facility is in place. This establishes the requirements and the amount of funds necessary to build a facility that will satisfy the product. The process can take much longer than the five-year minimum, depending upon the allocation of funds associated with the specific programme and its associated priorities.

In the case of overseas operations, the facility acquisition takes much longer. Because of the long acquisition cycle, the need for new facilities must be recognised early in the product's life cycle. If this need is not recognised, there is a serious problem. It must be determined early on what the gross facility requirements will be. Where these are deemed inadequate, new facility requirements should be developed. Particular difficulties in scheduling problems include approval, design and construction of the new facilities to support testing activities. They must be determined early in the concept phase if they will be available when required. Inputs to facility requirements planning include existing facility data, projected space availability, facility funding, constraints and projected operational maintenance concepts. Once these data are made available, they must be combined and analysed to determine the size of the additional space of the new facility that will be required to support the product. This effort should also take into consideration the size of the maintenance facility and the additional tools and test equipment that it will house.

Implementing changes to an existing facility will require a review of its present status. One must be in the position to determine the adequacy of the facility; specifically, concerning power, heating, ventilating, air conditioning, support services and present reliable space. Once this has been established, an analysis should be performed to determine what the new product requirements will be: how unique its power needs are, whether special rooms are needed to support the repair and servicing of the equipment, if tools and test equipment using unique metrology approaches to ensure the calibration of the equipment are available, if special lighting is needed, or whether clean rooms are required to ensure that the quality and workmanship of the equipment is retained at its original status. All of these help determine the new or improved facilities requirements base.

The logistician must ensure that all facilities are reviewed to ensure the handling capacity of the new product being distributed. One such concern is the training facility. Because of the advent of the computer in the computer-aided instruction base, the logistician should evaluate the availability of physical and computer space that can be allotted. If the computer-aided instruction equipment can accept additional documentation to support the present product or can adequately support minor changes to an existing curriculum new facilities may not be needed. Depending on the size and complexity of the equipment, one may have to consider additional computer capabilities to assist in the training of personnel required to operate and support it. This again may require additional rooms, more training capabilities, further power, heating, ventilating and air conditioning, better lighting and support facilities for the students. The flow of these students through the housing facilities can also become a problem because the greater the need for additional training, the more likely appropriate housing will not be available to support these personnel. All of this must be considered within the logistician's framework.

Facility planning

Facility planning is undertaken early in the programme development so that the appropriate space will be available to the end user or customer and he can be in a position to appropriate the necessary room to support either the operation or maintenance of the product that he will accept. Studies are started, as earlier indicated, as far back into the programme as possible. First, under the normal operational cycle, it takes time to build a facility and properly integrate it into the infrastructure already established within the accepting unit. Also, it takes time, money, and appropriate technical expertise to ensure that what is being built will satisfy the needs of the product. The following factors must be considered early in the programme:

- When the equipment is accepted, it must be integrated and properly interfaced into the equipment it may be supporting.
- If it is a stand-alone item, the appropriate power, air conditioning, heating, cooling and ventilation have been prepared and should be in place at time of acceptance of the new equipment.

Chapter 8 LOGISTICS SUPPORT ANALYSIS

Before proceeding to the actual discussion of logistics support analysis, it is appropriate to review the introduction of logistics so that the reader will have a good understanding of how LSA has become such a predominant factor in the generation of ILS. The ILS working group, as discussed in Chapter 1, had problems in being rapidly accepted by the engineering functions within an operating organisation. This also applies to the acceptability of the logistics people within the DOD as well as in the commercial environment. What was considered important was a mechanism or process that would assist in the integration of all of the engineering functions as well as those of all of the support and logistics functions. Logisticians recognised the need to repackage logistics if they were to be accepted by the engineer. This repackaged process required the information generated by the engineer to be quickly integrated by the logistics function, and the information generated by the logistician could be readily reviewed by the engineer. The logistics efforts entailed many analyses and trade-offs that only responded to the needs of the logistician. Many times, the analyses were not integrated into those of the design engineer. Therefore, a mechanism was required to close the loop between the two operating functions. For this to happen, it was decided that the logistics function had to appreciate the design engineer's intentions. That meant compromise from the logistics point of view. In relative terms, however, that compromise was not difficult, as the analytical approaches being used by the logisticians were no different from those being used by the engineers. Therefore, the fundamental concept of ILS is to relate support to design and to use an engineering analytical approach rather than a rule of thumb to design logistics support subsystems for the development of hardware to support acquisitions.

Using the definitions established by the DOD and defined in DOD Directive 5000.39, ILS is:

> a disciplined, unified and iterative process to the management of technical activities necessary to integrate support considerations into systems and equipment design, to develop support requirements that are related consistently to readiness objectives to design and to each other, to acquire the required support and to provide the required support during the operational phase at a mimimum total cost.[1]

This essentially said that all elements associated with logistics needed to be bonded together rather than being structured as independent elements. The elements discussed in Chapter 7 are the elements that logisticians must examine when assembling the LSA package. Mentioned in Chapter 2 and portrayed as part of Fig. 2.4 was an umbrella under which ILS operates. All of the interacting functions necessary to support the transfer of data among the relative aspects of the functioning areas within a programme are controlled by the programme manager. The concept that information relative to supporting the design becomes part of the LSA process applies here, as part of the ILS/LSA interface. It is entered into the database so that information relative to the actual design becomes common to both the logistician and the design engineer. Access to this material is available to each functional area, as well as all of the other functional areas that have inputted information into the programme.

It was discussed in Chapter 2 how the logistician and the programme manager interacted to support the specific programme in question. Fig. 8.1 puts this in quantitative terms. The elements specified are necessary to ensure that the ILS efforts are defined and measured during the conceptual design phase. Also specified are the detailed design and the logistics support planning. Together they have a direct effect on the readiness and measurement capabilities of the programme. These are the inputs in the quantitative analyses that the logistics engineer must perform so that the readiness and effectiveness measurement techniques are completed. How was this accomplished prior to LSA? In essence, they were being done independent of the design engineering group, and were isolated to the element manager responsible for the individual task. Feedback rarely affected the design, unless there was an aggressive logistics manager willing to discuss the pros and cons of the varying approaches with the design engineer. This was done to ensure that the logistics support aspects of the programme were being accomplished. Otherwise, it did not occur.

The question, then, was what were the differing aspects of logistics and design engineering, and how and when were they integrated during the

---

[1] US Department of Defense, *Acquisition and Management of Integrated Logistic Support for Systems and Equipment*, DOD Directive 5000.39.

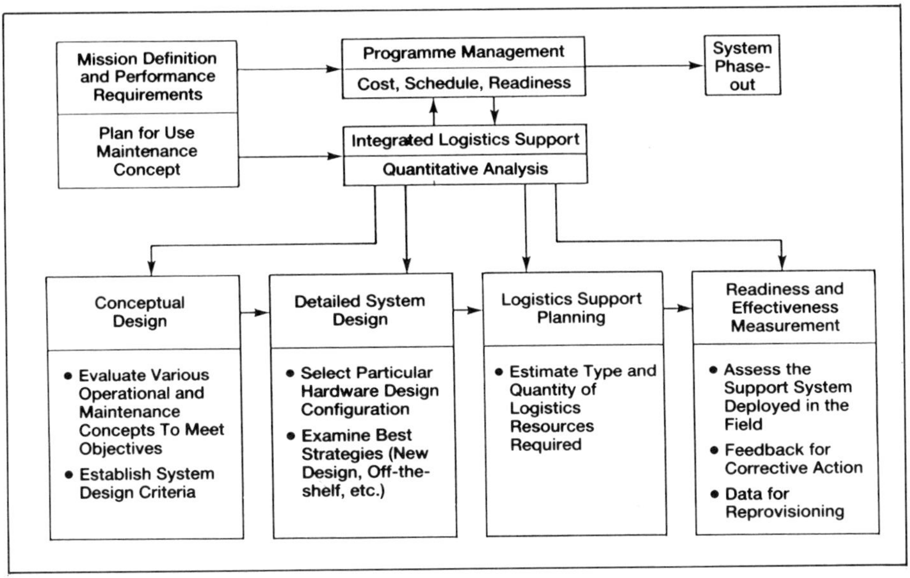

Fig. 8.1 ILS quantitative methods.

acquisition phase? Specifically, what was the relationship between design and support during the acquisition cycle of a programme? Table 8.1 portrays the efforts performed during the full life cycle of a product. The efforts of the design engineer and the logistician have been broadly described in the lists that accompany each of the acquisition phases. The lists are broad enough to provide a better appreciation of the prime efforts that must be performed during each phase of the contract life cycle.

To summarise, and perhaps to paraphrase some of the discussions presented as part of Chapter 2, the purpose of the ILS effort is:

- to ensure a proper planning process through the product life cycle;
- to establish a management discipline early enough in the design to ensure that the logistics and engineering groups accomplish what they set out to do;
- to establish requirements in analytical, quantitative and qualitative terms, so that a definitive approach can be tailored to satisfy the needs of the programme in question;
- to help establish a standardised approach to ILS in the application and use of the LSA.

The relationship of ILS and systems engineering to LSA

As indicated, the LSA is intended to be used as the integrating function between design engineering and the integrated logistics support group. A basic strategy to accomplish this was to identify the specific task outputs

Table 8.1 LSA tasks

Task section	*Purpose of task section*	*Task subtask*	*Influence*			*Applicability by phase*					*LSAR interface*
			System equipment design	*Support system design*	*Logistics requirements determination*	*Pre-concept*	*Concept*	*DVAL*	*FSD*	*Production deployment*	
100 – Programme planning and control	To provide for formal programme planning and review actions	101 – Development of an early logistic support analysis strategy	Primary Purpose of 100 Series Tasks is the management and control of the LSA programme			NA	G(3)	G(3)	S(3)	NA	NA
		101.2.1 – LSA strategy									
		101.2.2 – Updates									
		102 – Logistic support analysis plan				NA	G	G	G	NA	NA
		102.2.1 – LSA plan									
		102.2.2 – Updates									
		103 – Programme and design reviews				NA	G(2)	G(2)	G(2)	G(2)	NA
		103.2.1 – Establish review procedures									
		103.2.2 – Design reviews									
		103.2.3 – Programme reviews									
		103.2.4 – LSA review									
200 – Mission and support systems definition	To establish supportability objectives and supportability-related design goals, thresholds, and constraints through comparison with existing systems and analyses of supportability, cost and readiness drivers	201 – Use study				G	G	G	G	NA	Indirectly(4)
		201.2.1 – Supportability factors	X	X							
		201.2.2 – Quantitative factors	X	X							
		201.2.3 – Field visits	X	X							
		201.2.4 – Use study report and updates	X	X							
		202 – Mission Hardware, software and support system standardisation				NA	G(2)	G(2)	G(2)	G(2)	Indirectly(4)
		202.2.1 – Supportability constraints	X	X	X						
		202.2.2 – Supportability characteristics	X								
		202.2.3 – Recommended approaches	X	X							
		202.2.4 – Risks	X	X	X						
		203 – Comparative analysis				G	G	G	G	NA	Indirectly(4)
		203.2.1 – Identify comparative systems	X	X		G	G	G	NA		
		203.2.2 – Baseline comparison system	X	X		G	G	G	G	NA	
		203.2.3 – Comparative system characteristics	X	X		G	G	G	NA	NA	
		203.2.4 – Qualitative supportability analysis	X	X			G	G	G	NA	
		203.2.5 – Supportability, cost and readiness driver	X	X		G	G	G	G	NA	
		203.2.6 – Unique system drivers	X	X		G	G	G	NA	NA	
		203.2.7 – Updates	X	X		NA	G	G	G	NA	
		203.2.8 – Risks and assumptions	X	X		G	G	G	G	NA	
		204 – Technological opportunities				NA	G	G	S	NA	Indirectly(4)
		204.2.1 – Recommended design objectives	X	X							
		204.2.2 – Updates	X	X							
		204.2.3 – Risks	X	X							
		205 – Supportability and supportability-related design factors		X		NA	G	G	G	NA	A record
		205.2.1 – Supportability characteristics				NA	G	G	NA	NA	
		205.2.2 – Supportability objectives and associated risks	X	X		NA	G	G	NA	NA	
		205.2.3 – Specification requirements	X	X		NA	G	G	G	G	
		205.2.4 – NATO constraints	X	X		NA	G	G	NA	NA	
		205.2.5 – Supportability goals and thresholds	X	X		NA	NA	G	NA	NA	

			Influence			*Applicability by phase*					
Task section	*Purpose of task section*	*Task subtask*	*System equipment design*	*Support system design*	*Logistics requirements determination*	*Pre-concept*	*Concept*	*DVAL*	*FSD*	*Production deployment*	*LSAR interface*
300 – Preparation and evaluation of alternatives	To optimise the support system for the new item and to develop a system which achieves the best balance between cost, schedule, performance and supportability	301 – Functional requirements identification				NA	G	G	G	C	
		301.2.1 – Functional requirements		X		NA	G	G	S(1)	C(1)	C Record
		301.2.2 – Unique functional requirements		X		NA	G	G	S(1)	C(1)	C Record
		301.2.3 – Risks		X		NA	G	G	S(1)	C(1)	
		301.2.4 – Operations and maintenance tasks		X	X	NA	S	G	G	C	B/C/D
		301.2.5 – Design alternatives	X			NA	G	G	G	C	Records
		301.2.6 – Updates		X	X	NA	G	G	G	C	
		302 – Support system alternatives				NA	G	G	G	C(1)	
		302.2.1 – Alternative support concepts		X		NA	G	G	NA	NA	
		302.2.2 – Support concept updates		X		NA	G	G	S	NA	
		302.2.3 – Alternative support plans		X		NA	S	S	G	C(1)	Indirectly(4)
		302.2.4 – Support plan updates		X		NA	S	S	G	C(1)	
		302.2.5 – Risks		X		NA	G	G	G	C(1)	
		303 – Evaluation of alternatives and trade-off analysis				NA	G	G	G	C	
		303.2.1 – Trade-off criteria	X	X	X	NA	G	G	G	C	
		303.2.2 – Support system trade-offs		X		NA	G	G	G	C	E/F/G/
		303.2.3 – System trade-offs	X	X		NA	G	G	G	C	Records
		303.2.4 – Readiness sensitivities	X	X	X	NA	G	G	G	NA	
		303.2.5 – Manpower and personnel trade-offs	X	X	X	NA	G	G	G	NA	
		303.2.6 – Training trade-offs	X	X	X	NA	G	G	G	C	
		303.2.7 – Repair level analyses		X		NA	S(1)	G	G	C	C Record
		303.2.8 – Diagnostic trade-offs	X	X		NA	G	G	S(1)	NA	
		303.2.9 – Comparative evaluations		X	X	NA	G	G	S(1)	C	
		303.2.10 – Energy trade-offs	X	X		NA	G	G	S	C	
		303.2.11 – Survivability trade-offs	X	X		NA	G	G	G	C	
		303.2.12 – Transportability trade-offs	X	X		NA	G	G	NA	NA	
400 – Determination of logistics support resource requirements	To identify the logistics support resource requirements of the new system in its operational environment(s) and to develop plans for post-production support	401 – Task analysis				NA	NA	S	G	C	
		401.2.1 – Task analysis			X	NA	NA	S	G	C	C/D/D1
		401.2.2 – Analysis documentation			X	NA	NA	S	G	C	record
		401.2.3 – New/critical support resources			X	NA	NA	S	G	C	E/F/G
		401.2.4 – Training requirements and recommendations			X	NA	S	G	G	C	records
		401.2.5 – Design improvements	X	X		NA	NA	S	G	C	D1 record
		401.2.6 – Management plans			X	NA	NA	S	G	C	J record
		401.2.7 – Transportability analysis	X	X		NA	NA	S	S(1)	C(1)	H/H1
		401.2.8 – Provisioning requirements			X	NA	NA	G	G	C	records
		401.2.9 – Validation	X	X	X	NA	NA	S	G	C	All record(s)
		401.2.10 – ILS output products			X	NA	NA	S	G	C	All record(s)
		401.2.11 – LSAR updates	X	X	X	NA	NA	S	G	C	
		402 – Early field analysis				NA	NA	NA	G	C	
		402.2.1 – New system impact			X						
		402.2.2 – Sources of manpower and personnel skills			X						
		402.2.3 – Impact of resource shortfalls			X						
		402.2.4 – Combat resource requirements			X						
		402.2.5 – Plans for problem resolution			X						
		403 – Post-production support analysis				NA	NA	NA	NA	G	
		403.2 – Post-production support plan		X	X						
500 – Supportability assessment	To ensure that specified requirements are achieved and deficiencies corrected	501 – Supportability test evaluation and verification	X	X	X	NA	G	G	G	G	
		501.2.1 – Test and evaluation strategy	X	X	X	NA	G	G	S	NA	
		501.2.2 – Objectives and criteria	X	X	X	NA	NA	G	G		
		501.2.3 – Updates and corrective actions	X	X	X	NA	NA	G	G	S	All record(s)
		501.2.4 – Supportability assessment plan (post-deployment)	X	X	X	NA	NA	NA	G	S	
		501.2.5 – Supportability assessment post-deployment	X	X	X	NA	NA	NA	NA	G	

Table 8.1 (continued). Footnotes

Programme phases are characterised by the following design status:

1. PRE-CONCEPT – No design. Mission area analyses are performed on a continuing basis to include supportability and sustainability considerations within mission areas. Programme requirements grow out of these analyses.
2. CONCEPT – Design is only conceptual. Best opportunity for identifying alternatives, conducting trade-offs, and influencing design from a supportability standpoint.
3. DVAL – Performance characteristics are more or less established. Actual design is still flexible. Debugging and major changes in construction are taking place. Support alternatives and support, design, and operations alternatives are being traded. May result in a prototype.
4. FSD – Results in a prototype. Design is concentrating on construction, parts selection, and fine tuning of performance. No major design influence is possible. Design influence is limited to packaging, partitioning, testability, accessibility, etc. Support system is optimised.
5. PROD – Design is fixed. Logistic support resource planning is complete. No opportunity for trade-offs or further optimisation.

Applicability by phase code definition: C, generally applicable to design changes only; G, generally applicable; NA, not applicable; S, selectively applicable.

(1) – Requires considerable interpretation of intent to be cost effective.
(2) – MIL-STD-1388-1A is not the primary implementation document. Other MIL-STD's or statement of work requirements must be included to define the total requirements.
(3) – Done just prior to initiation of the phase.
(4) – Selectively applicable for equipment level acquisitions.
(5) – Not applicable for equipment level acquisitions.

and the projected task requirements that were going to be performed to ensure that the LSA process and the integration of engineering and ILS are accomplished. Fig. 8.2 is a portrayal of what will be discussed in the following text. The actual engine associated with the LSA is the centre point of the diagram. The plumbing that supports the engine represents the efforts required so that the engine can actually work. Fuel is added to the engine via the existing support requirements, which appear at the top of the logistics support analysis machine, and via the design engineering process. Both of these elements are equal in scope and relationship to the total systems approach. The result of the work being performed by the engineering process and the logistics support process is a very supportable product defined by an equally important and compatible system support requirement operating simultaneously, and in parallel. Logistics support analysis is intended to be the 'I' in integrated logistics support, to be the integrator so that all of the support functions and the design aspects can be properly organised and orchestrated. This effort can only be accomplished if the programme manager properly orchestrates the work that needs to be performed, the efforts that need to be traded off and the specific communications networks that need to be established to ensure that he has brought to his attention and appropriately highlighted the efforts that need to be resolved at his management level. The items that can be traded off at a lower level (specifically, between the functions) should be accomplished and the records kept via the data record base.

What is LSA?

The LSA process is the engineering process applied to a product being produced. Its objective is to achieve a balance between readiness, operational capability and cost, and the systems logistics requirements. This is a multiple objective, from the standpoint of being able to balance the tradeoffs between each of the elements associated with design and those that are output through the LSA to ensure a compatible logistics support system. It therefore becomes a tool to integrate ILS with the other engineering functions to ensure that the system design and operational requirements have been properly applied via a single analytical approach. This ensures that the logistics support requirements have been satisfactorily addressed and accomplished. It is also used to eliminate the barriers among all of the other operating functions so that each of these people would not be isolated, but would attempt to work with each other through local area networks applying the techniques of the LSA to help meet the requirements of the contract. Fig. 8.3 shows what occurred prior to LSA. Some of the walls have been taken down, and some of the prejudices have gone away, but the logistician should stick to working closely with the other functional elements of the organisation to ensure that the LSA process does what it is intended to do. The LSA process is not a panacea, but an additional tool to be used by the ILS manager and the other functional managers to assist in defining and refining a design.

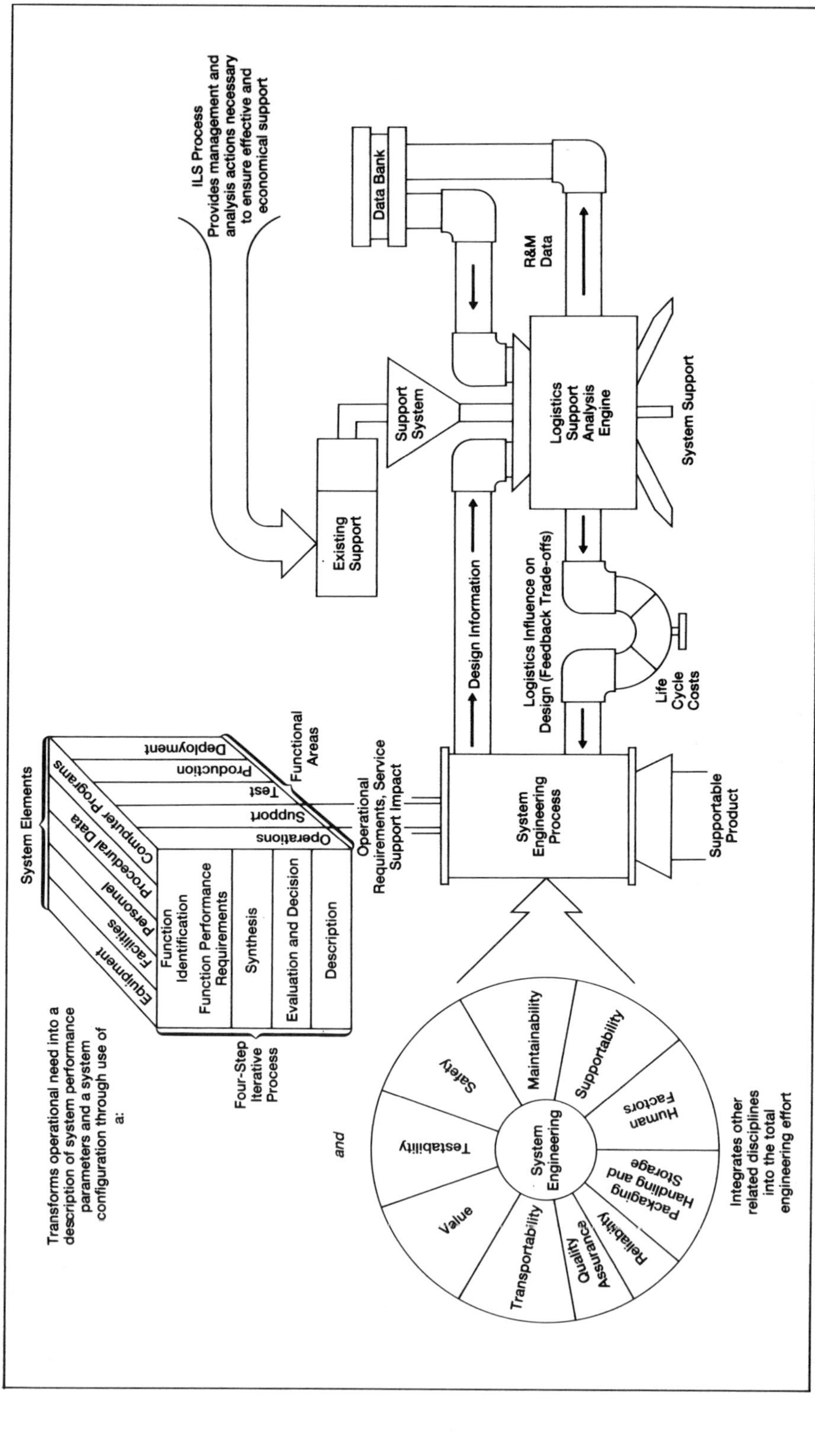

Fig. 8.2 The relationship of ILS and systems engineering to logistics support analysis, PH&S, package handling and storage.

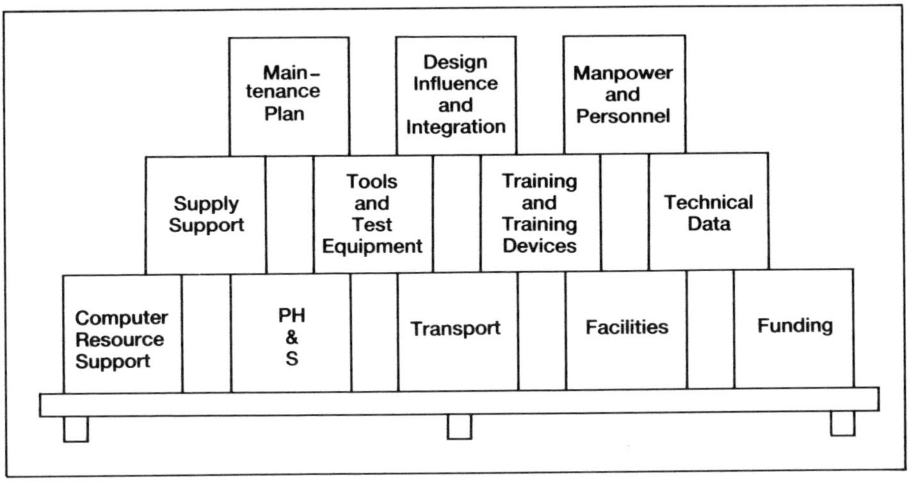

Fig. 8.2 — Contd.

Fig. 8.3 'Working in a closet'.

Why LSA?

The basic requirement is the application of scientific and engineering efforts to enable support considerations to influence the design quickly. This problem can be avoided if the logistician is able to talk in engineering terms with the engineer and appropriate trade-off studies and analyses can be performed. Thus, any influence on the design, based upon the support requirements, can be made. LSA is also used to help identify cost drivers and support problems early in the design phase. Entering appropriate data into the LSA database as early as possible will help identify the possible design problems as they relate to logistics support. The specific cost drivers are also quickly analysed and measured to determine how best to address those specific elements. Normally, measurement is made of such efforts as reliability and its appropriate costs, maintainability, design to satisfy requirements and the integration of specific diagnostic routines and how best to ensure that the diagnostics that are applied satisfy the minimum essential requirements of the contract. These three elements could drive up the cost of the design beyond limits so that it becomes extremely unrealistic.

The LSA is also used to help develop projections of the logistics support resources needed to support the new product over its expected life cycle. What does this really mean? The question really is how many more people are going to be required to support the design once it is distributed? What additional facilities will be required if the device fails more often than anticipated? What new test equipment will be required to troubleshoot the equipment in question? All of these questions can be quickly reviewed and problems avoided if these logistics projections can be reasonably estimated prior to entering into the full-scale engineering development. What does LSA actually do? It helps establish a single database from which all of the requisite material to support logistics can be derived. It uses the engineering process to ensure that all of the aspects associated with logistics and those being inputed by design are properly evaluated and measured. This is done so that a design support programme can be sufficiently described to accomplish the logistics support goal.

Before proceeding further, it is important to note that this book does not go into the specific details of how the LSA process is to be performed, what elements should be performed, and the details of any performance. Rather, the intent is to give the reader an overview of the LSA and its appropriate processes, identifying in higher level terms the efforts that should be performed during the different phases of a contract and what the results should be. The specific aspects of how to perform an LSA, what detailed information should go into it, what the individual sheets are, what they are composed of and how to fill them out is available in other reading material.

The ILS/LSA process

LSA tailoring

The LSA should not be fully invoked on programmes that are defined either by the prime contractor or his related subcontractor, or implemented by the acquisition manager in the procuring activity to the prime contractor. The LSA is intended to satisfy the absolute minimum requirements. If these objectives are not accomplished, the LSA will not be cost effective, nor will it be favourable to the people who have to perform and evaluate it. It is therefore recommended that if the prime contractor is going to implement LSA, he should talk separately with each of the vendors or subcontractors. The prime contractor should generate an approach to best meet the specific requirements levied as part of the prime contract. Not all tasks need to be performed, especially if the vendor is supplying off-the-shelf equipment. In off-the-shelf equipment the design process is completed, and the need to affect the specific product in question is essentially eliminated. Only when there is a major design process or major modifications to an existing design should LSA be invoked. The data associated with goals like reliability and maintainability should be requested and supplied by the source rather than having all of the data submitted in LSA format. Compatibility and ease of implementation is important.

The procuring activity also needs to review this requirement before adding the requirements of LSA to the prime contractor. Absolute tailoring is essential, ensuring minimum requirements are achieved to optimise the application of LSA. Whether it is done in an automatic, semi-automatic or a manual fashion, the request for data should be carefully scrutinised so that only the minimum material is delivered to support the product in question. To support this approach some of the considerations that should be addressed in the tailoring process both by the procuring activities as well as the prime contractor, are:

- The type of programme and where it is in the development stage, whether it is a product improvement or development of a new product.
- The amount of design freedom available to the specific product in question.
- Time and resources available to accomplish the tasks.
- The work that has already been done (there is no need to duplicate this existing effort).
- Past experience in historical data.
- The acquisition strategy assigned to the programme by the procuring activity and/or the prime contractor.

Additional guidance on these specific factors is provided in MIL-STD 1388. Fig. 8.4 shows a general tailoring logic tree that should be followed in selecting the specific tasks appropriate for the design programme that is being considered.

The management functions associated with the LSA process

The LSA process resides within the logistics manager's responsibilities. The logistics manager cannot do all of the work so he should delegate responsibility for the LSA to an individual who has the knowledge and capability to integrate the process between engineering and logistics. The

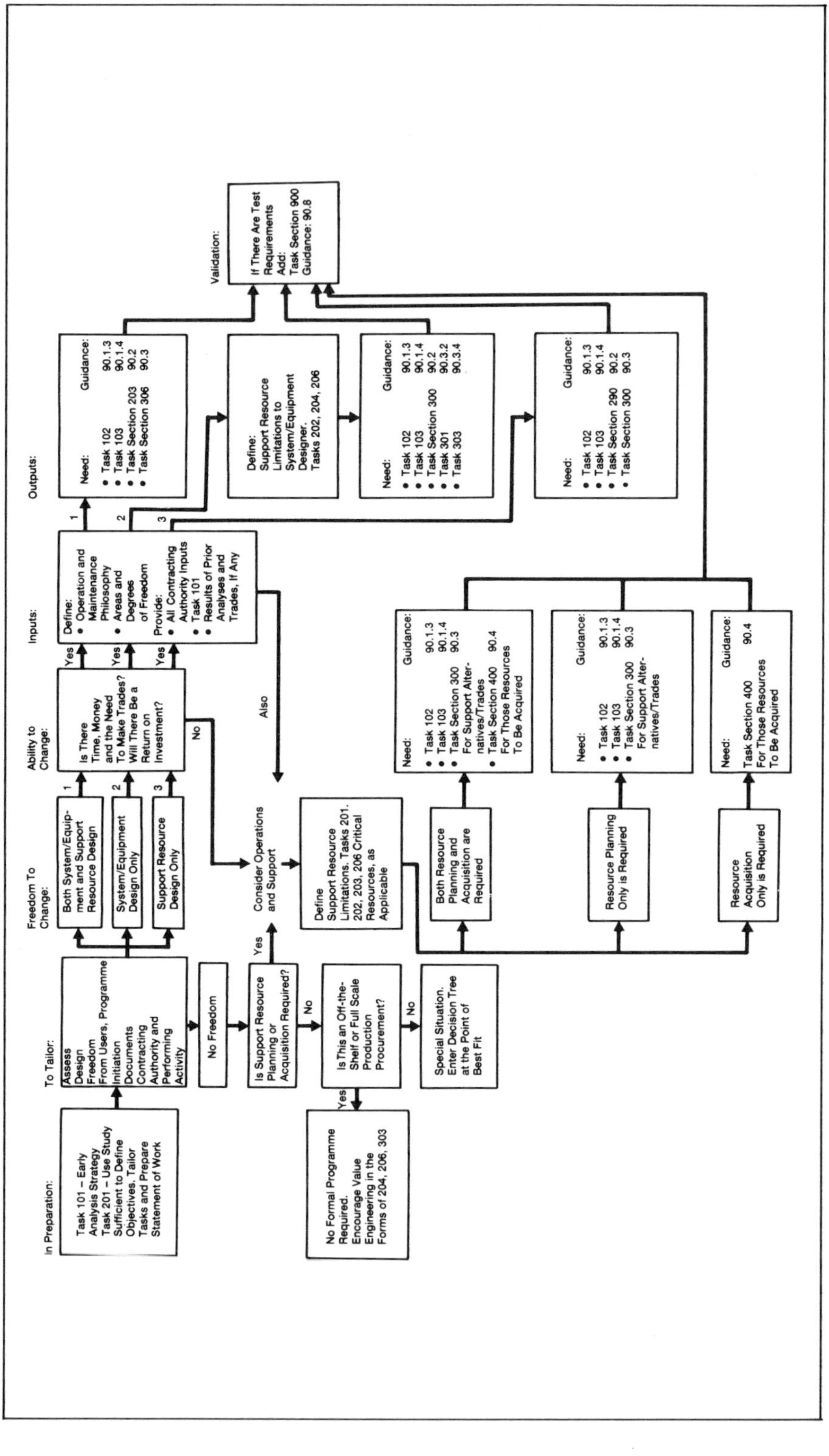

Fig. 8.4 Logistics support analysis: tailoring decision logic.

manager of the LSA should be responsible to the ILS manager, just as the logistics element managers are responsible to him for the performance of their specific tasks. Four specific elements are involved in managing the LSA process:

1. Planning, which requires the identification of LSA objectives and schedules and the actions required to achieve them. Planning is also decision making, as it involves selective alternative approaches in the specific designs in question.
2. Organising, which involves establishing an LSA organisational structure for the receipt of specific data coming in from the logistics elements managers as well as from the engineering functions that identify the LSA tasks and subtasks, assigns tasks to the specific organisational units and delegates the appropriate authority to them. It also coordinates authority relationships, both within the LSA programme structure and within its related programme efforts.
3. Staffing involves manning and keeping manned the technical and management positions identified in the organisational structure. This falls under the operational role of the LSA managers but it is the direct responsibility of the ILS manager.
4. Directing also requires the need to provide motivation, communication and leadership. The chain of events requires the crossing of many lines of authority and responsibility to ensure that the LSA process is in fact accomplished and that the controlling aspects (which means the measurement of performance) correct any deviations and ensure accomplishment of the plans. The organisational management structure associated with the LSA becomes part of the ILS training documentation that was discussed in Chapter 2. It becomes the operating procedures by which the LSA manager operates and establishes guidelines for each of the interfacing organisations related to the LSA efforts.

Under the management functions, good coordination solves problems as they occur; excellent coordination anticipates and prevents problems. Effective coordination of the LSA programme with its related elements will identify mutual interests, maximise the benefits of mutually supporting tasks and prevent duplication of efforts. The manager should be dynamic and have a full understanding of the impacts on both logistics and engineering of the decisions that need to be made to support the programme.

The logistics support analysis record

The logistics support analysis record (LSAR) is in essence a summarised logistics database from which all logistics support work will be performed. It provides a tool for the procuring activity as well as the contractor's team so that it integrates everybody's information. It becomes accessible to all of the

Table 8.2

LSAR Reports	LSAR Data Records											
	A	B	B1	B2	C	D	D1	E	F	G	H	H1
LSA–001					•		•					
LSA–002					•		•					
LSA–003	•											
LSA–004					•		•					
LSA–005					•		•					
LSA–006					•		•					
LSA–007					•		•					
LSA–008					•		•					
LSA–009											•	•
LSA–010												
LSA–011					•		•					
LSA–012					•		•					
LSA–013					•		•					
LSA–014					•		•					
LSA–015					•	•	•					
LSA–016					•		•				•	•
LSA–017					•		•					
LSA–018											•	•
LSA–019					•		•					
LSA–020					•		•				•	
LSA–021					•	•						
LSA–022					•	•						
LSA–023	•	•		•	•		•				•	•
LSA–024	•	•			•		•				•	•
LSA–025											•	•
LSA–026											•	•
LSA–027	•			•	•							•
LSA–028											•	•
LSA–029											•	•
LSA–030											•	•
LSA–031											•	•
LSA–032											•	•
LSA–036											•	•
LSA–040											•	•
LSA–041											•	•
LSA–042											•	•
LSA–043											•	•
LSA–050		•	•	•								
LSA–051		•	•	•								
LSA–052		•	•	•								
LSA–053		•	•	•								
LSA–054		•	•	•								
LSA–055		•	•	•								
LSA–151											•	

Key for table 8.2

LSAR data records

Data record A	Operation and maintenance requirements
Data record B	Item reliability (R) and maintainability (M) characteristics
Data record B1	Failure modes and effects analysis
Data record B2	Criticality and maintainability analysis
Data record C	Operation and maintenance task summary
Data record D	Operation and maintenance task analysis
Data record D1	Personnel and support requirements
Data record E	Support equipment or training material description and justification
Data record E1	Unit under test and automatic test programme(s) description
Data record F	Facility description and justification
Data record G	Skill evaluation and justification
Data record H	Support items identification
Data record H1	Support items identification (application-related)
Data record J	Transportability engineering characteristics

LSAR reports

LSA-001	Direct Annual Maintenance Man-Hours by Skill Speciality Code and Category of Maintenance
LSA-002	Personnel and Skill Summary
LSA-003	Maintenance Summary
LSA-004	Maintenance Allocation Summary
LSA-005	Support Item Utilisation Summary
LSA-006	Critical Maintenance Task Summary
LSA-007	Support Equipment Requirements
LSA-008	Support Items Validation
LSA-009	Support Items List
LSA-010	Parts Standardisation Summary
LSA-011	Requirements for Special Training Devices
LSA-012	Requirements for Facilities
LSA-013	Support Equipment Grouping Number Utilisation
LSA-014	Training Task List/Task Inventory List
LSA-015	Sequential Task Description
LSA-016	Preliminary Maintenance Allocation Summary
LSA-017	Preliminary Maintenance Allocation Summary Tool Page
LSA-018	Visibility and Management of Operating and Support Cost (VAMOSC)
LSA-019	Maintenance Task Analysis Validation
LSA-020	Tool and Test Equipment Requirements
LSA-021	Task Reference List
LSA-022	Referenced Task List
LSA-023	Maintenance Plan Summary
LSA-024	Maintenance Plan
LSA-025	Packaging Requirements Data
LSA-026	Packaging Developmental Data
LSA-027	Failure/Maintenance Rate
LSA-028	Reference Number/Additional Reference Number Cross Reference List
LSA-029	Repair Parts List
LSA-030	Special Tools List
LSA-031	National Stock Number and Part Number Index
LSA-032	DLSC Submittals
LSA-036	Provisioning Requirements
LSA-040	Components of End Item (COEI)
LSA-041	Basic Issue Items (BII)
LSA-042	Additional Authorisation List (AAL)
LSA-043	Expendable/Durable Supplies and Materials List (ESML)
LSA-050	Reliability Centred Maintenance Summary
LSA-051	Reliability Summary/Redesign
LSA-052	Criticality Analysis
LSA-053	Maintainability Analysis Summary – Level of Repair
LSA-054	Failure Mode Analysis Summary
LSA-055	Failure Mode Detection Summary
LSA-060	LSA Control Number Master File
LSA-061	Parts Master File
LSA-080	Bill of Materials
LSA-100	Chronolog Information
LSA-101	Transaction Edit Results – Selection Cards
LSA-102	Transaction Edit Results – LCN Master
LSA-103	Transaction Edit Results – Parts Master
LSA-104	Transaction Edit Results – Narrative Master
LSA-105	Key Field Change Transactions
LSA-106	Reference Number Discrepancy List
LSA-107	LCN/Task Identification Code Cross Reference List
LSA-108	Critical Data Changes
LSA-109	Unidentified Transactions
LSA-150	Provisioning Error List
LSA-151	Provisioning Parts List Index (PPLI)
LSA-152	PLISN Assignment/Reassignment

people required to interface with the LSA process and it gives the basis from which all of the associated and supporting data are derived. The information becomes usable and manageable so that all of the personnel responsible have a standard database from which to draw the information required to support any future analysis. Lastly, it creates the audit trail necessary to support any of the changes being proposed or incorporated into the design. It also establishes the reasoning for not implementing a specific change, and it provides the support data for the changes that were implemented. The LSAR therefore provides a corporate memory of the design and logistics support databases.

Purpose of the LSAR

The LSAR is used to provide the medium for recording the logistics support analysis data. It is the basis by which all of the information is compiled. It serves as the source data for supporting documentation, the outputs becoming necessary documentation to ensure that the operational personnel have the appropriate material available to them when the equipment is distributed or deployed. It also provides the database for modifications and improvements necessary to support the programme and it is used for comparative analysis based upon the results coming in from the field. This becomes the historical data for the specific equipment and it is used as a basis to compare the existing support system to any new or revised support that would be required for an up and coming product.

Table 8.2 shows what the data sheets represent, how they should be employed and the scope and appropriate title. These sheets are intended to assist in the summarisation of the specific LSAR requirements and are directly related to the elements specified in MIL-std-1388-2A. Again these are used only for summation. More definitive information can be obtained from MIL-std-1388-2A, which provides all the necessary information, application and approaches to completing the data sheets.

Chapter 9 LIFE CYCLE COSTING

Introduction

Life cycle costing (LCC) has been designed to establish the baseline cost for a product being developed. It is primarily used as a decision-making tool during the concept design and acquisition phase. LCC requires the identification of all potential costs and the means for predicting these costs over time. At present LCC has tended to become an artifice for the accumulation of cost according to some cost accounting format, often called the work breakdown structure, which requires the identification of cost to a precise measurement and allows for no future uncertainties. Cost accounting does not permit LCC to be used for design trade-off decisions. LCC is more than an automatic ritual of computing the cost as defined in the textbooks or summing up costs per the cost breakdown structure. Instead, LCC is essential to support decisions. It should become part of the overall engineering effort and should be supported by the relevant 'user oriented' cost concept.

Life cycle costing should be one of the most dynamic tools used by logistics and engineering management to assess the progress that is being achieved during all phases of design.

Very few decisions are made during the programme's life cycle that do not affect LCC. Programmatic design choices can cause variance in LCC and have a significant impact on the product's effectiveness. The use of LCC is most effective during the early phases of an acquisition cycle. By the time the concept definition is validated, approximately 85% of the product's life cycle has been committed via design and the choices made by the logistician. In real terms, Fig. 9.1 shows that at the completion and development of the prototype plans, the cost has been achieved. It is quickly recognised that the sooner one initiates an LCC modelling tool, the faster it will have an impact on the total logistics life cycle. What is being sought is the impact that the design changes will have on the product once it is used. (This is where the costs start escalating, not during the design phase). Clearly, the systems with the greatest chance of affecting the LCC and identifying subsequent savings are those that actually have an impact on acquisition and support (see Fig. 9.2). These occur during the preconcept and concept definition phase.

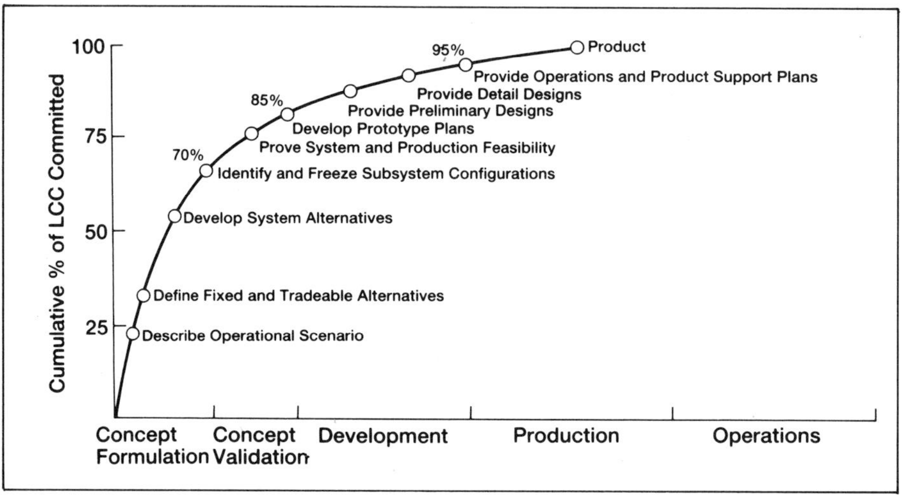

Fig. 9.1 Actions affecting life cycle cost.

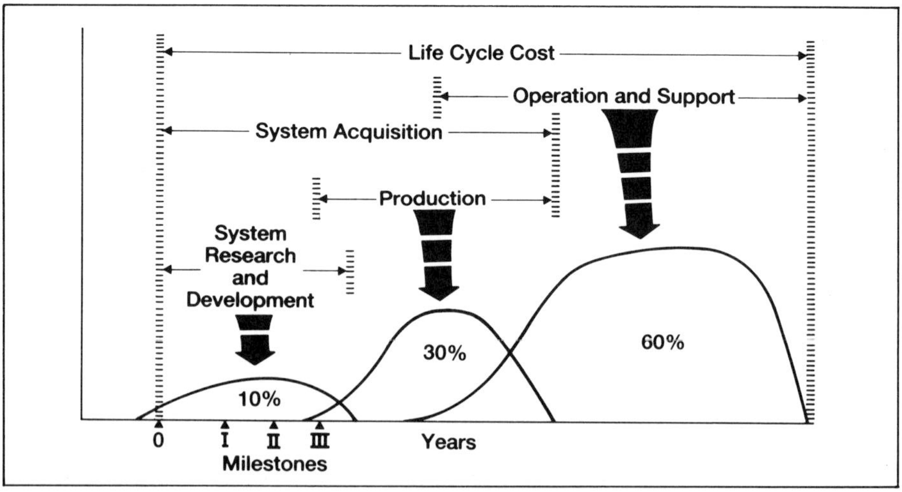

Fig. 9.2 Nominal cost distribution of a typical programme.

The goals of LCC are as follows;

- to identify the total cost of alternative means of solving a problem and achieving production schedules;
- to attain product performance and readiness objectives;
- to estimate the cost impact of the various design and support options.

Historically, a low initial acquisition cost for a specific product does not necessarily ensure a low life cycle support cost; in practice a low acquisition cost will more than likely be expensive. The author's belief is that life cycle cost should be done early and practised continuously. Thus, as decisions are made, appropriate sensitivity analysis can be performed using the LCC to determine what impacts the alternative approaches will have on the total system. It is not uncommon to find that the acquisition component of the programme may be increased to reduce support costs. It is paramount that the logistician work closely with the design engineer and with the cost analyst in determining what approaches should be performed. It also is imperative that as these alternative approaches are being considered, they are run through the model to determine what impact they will have on the total life cycle cost of the product.

Therefore, the following four factors affect LCC application. There is no attempt to establish policies or to dictate the means of resolving the problems of how and by whom LCC is applied.

- The general way in which a development is planned and structured will affect the environment for LCC and the possibility of utilising the LCC as an everyday decision-making tool.
- LCC will be affected by investment costs and investment opportunities.
- The scope of the decision strongly affects the relevant costs. Two concerns are to trace costs to the corresponding system effectiveness and the trade-off costs' elements versus cost elements and cost elements versus effectiveness.
- When summing up future life cycle costs, not only are there problems with the true value of money, but costs are subject to different degrees of uncertainties relative to time and the net effect on the product. This effort must be taken into account, especially when trying to predict costs.

The environment for decision making

To establish a policy stating that the decisions and context of engineering and logistics should be supported by user-oriented cost is one thing, but to have the policy implemented in an engineering management environment is something completely different.

It is quite evident that LCC, especially those related to support readiness, play a major role in the choice between two major configurations. Still, the analysis on which the decisions are based is generally performed because of favourable circumstances rather than prior planning. What should preferably be done is to perform the analysis because it is inherent in the product development planning.

In evaluating LCC use, the following list identifies some problems LCC has as an effective decision-making tool.

- Engineering management planning often does not include an efficient procedure by which choices between alternatives affecting LCC can be quickly identified.
- Most decisions are not clearly defined, if at all, regarding time factors, alternatives, evaluation criteria and the other overall scopes of development.
- Few decisions will have sufficient lead time to allow for LCC analysis.
- Few decisions based on one decision point will justify the cost of the analysis needed. The basic maintenance and support analysis will often lack timeliness for the single decision point and the given lead time.

With such problems in mind, it is obvious that the structure and the overall planning control in charge of product development create an environment for LCC that can vary considerably. To establish a basis for this hypothesis, two cases, the extreme and the norm, will serve to define a range of alternatives for a satisfactory LCC environment. Trade-offs between these extremes will add valuable information towards developing a viable LCC within a logistics and engineering management plan.

Task planning

In an extreme case, only the major milestones are defined as critical decision points. The following tasks are then derived from these regardless of cost and impact on sublevel decisions:

- define the scope of the programme;
- establish the subtasks;
- establish the time constraints on each task and the interrelationship between tasks.

The structure of the process is in terms of task times and resources. For obvious reasons, this approach might easily produce an unfavourable LCC environment. A prerequisite for LCC to become efficient in the decision-making role is to introduce ILS.

The other decision planning approach, the norm, accounts for both major and minor programme decisions. These decisions are identified in advance in a highly interrelated decision network. The steps with the norm are as follows:

- define the scope of the problem;
- idenfity system functions, including maintenance and support;
- identify decisions to be made to define all system functions;
- identify a decision breakdown corresponding to a function breakdown;
- identify time constraints and interrelationships between decisions.

The normal procedure, if applied to the fullest extent possible, will provide an excellent LCC environment. It should be noted that the scope of integration will follow from a policy that uses LCC. Under a task planning scheme, ILS is a necessary prerequisite for LCC.

Time factors

A decision point can represent a division between past and future costs. Past costs are generally considered investment costs. The characteristic of these costs is that, irrespective of which alternative is chosen, investment costs will be affected. As a consequence of this, and in accordance with accepted practice, such costs cannot be considered when evaluating alternatives and are therefore left out of the LCC decision. It can also be concluded that an LCC will always change with time. Investment costs will be continuously accumulated, and hence they will be excluded as relevant costs. The difference between an investment and a relevant life cycle cost may not be obvious from a statement to perform a cost analysis. It is therefore advisable to go explicitly to the analysis and illustrate this. Sometimes the use of LCC cost differences between alternatives can help to avoid these kinds of problems. A decision point can always represent the division between past and future decisions. A number of engineering decision schedules for the future represents a positive value in a situation where these costs must be made immediately. Since more information will eventually be available, it would be expected that a better choice could be made later by choosing a specific alternative. Present limits may be imposed on the range of alternatives at one of the forthcoming decision points. What has happened is the creation of investment opportunities, which can be rather costly.

In any analysis, assumptions must be made about future alternatives (for example, a support concept or repairability of a module). If the present decision depends too much on such assumptions, a decision based on that analysis might imply investment opportunities. Such problems are inherent in most LCC applications.

In an attempt to document all of the decisions made and why they were chosen, a flow diagram called Design Option Decision Tree (DODT) should be used. The DODT allows the analyst to present the trade-off analysis study pictorially. The DODT also illustrates the various options available during the design phase, so that when reaching decision points in the analysis, logical paths can be presented. The impact on future designs and decisions should therefore be reduced. The DODT will also restructure a set of cost elements in such a way that cost related to effective alternatives can be isolated. To support the presence of more than one alternative, sensitivity analysis should be performed so that any early changes will have minimal effects on future decisions. These DODTs allow the analyst to revise the investment opportunities and, if errors are caught early enough in the system, permit redirection to avoid costly errors that have been forecast.

It is also common to utilise the tools that are currently at hand. It is also more common to try to establish those tools so that they are integrated as part of the total design package. Therefore, the application of LSA in its integral applications of life-cycle cost models should be immediately applied to support the front-end design of a product.

The importance of running LCCs and the impact it will have on the design if it is appropriately applied has been discussed. The question really comes down to what should be done and what elements have a direct impact on the LCC when it is used to support the product in question. Fig. 9.3 illustrates the flow that will be required to ensure that the life cycle cost model and all of its elements are properly applied and determined so that there is a constant feedback to the engineer. Once a product has been defined, there should be some way of evaluating how the product is going to be utilised. The centre portion of Fig. 9.3 combines all of the models necessary to ensure that an appropriate life-cycle cost evaluation is being performed. Block 1 is part of the overall evaluation in the market analysis. This input data tells us first, from a commercial standpoint, how much a product should cost so that it will meet the consumer's needs. Product block 10, which is on the far side of the central portion, is what it will cost to repair it once it has failed. These take into consideration the levels of repair for a product.

Regardless of the market in which one is attempting to distribute a product, all elements and all models in the central portion of this system evaluation block diagram should be performed. The intent is to tailor the product to the market in which the product will be sold. From a military standpoint, the product distribution model would be more applicable than the market analysis model. The former, in essence, defines where and how many military devices will be stored. All of the other aspects are directly applicable to both the military and the commercial environment. The output of this process are the determining factors that will help access and evaluate specific impacts and determine how to tailor, analyse and minimise the different cost-related factors associated with some of the elements that are considered cost drivers.

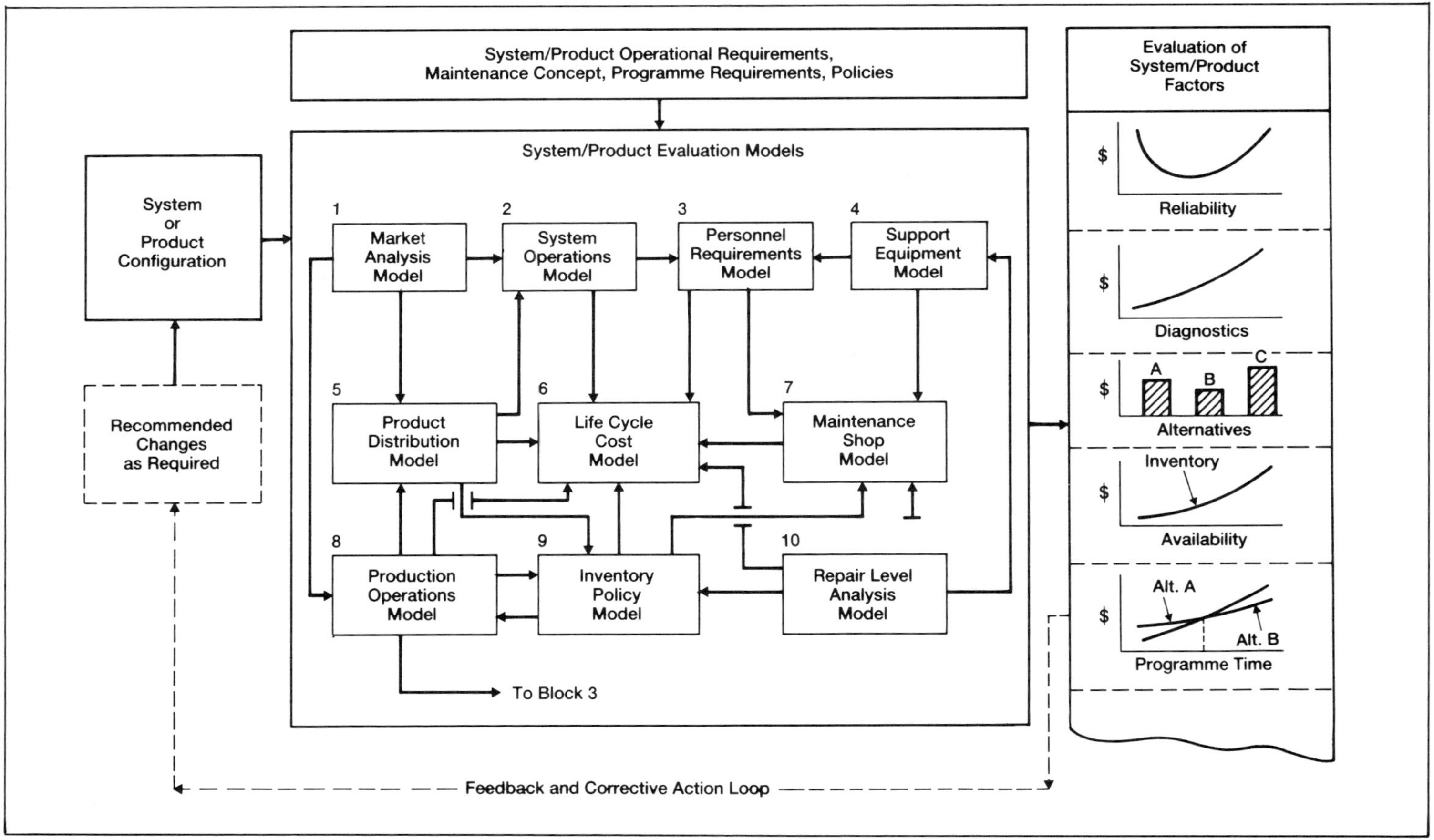

Fig. 9.3 Application of life cycle cost models.

In an attempt to illustrate the various evaluations performed to minimise the total cost of ownership, a series of graphs shown in Figs 9.4 – 9.8 exemplify this specific need. These figures cover elements such as personnel cost versus complexity, supply cost versus design standardisation, equipment accessibility versus cost and repair levels versus cost. These elements show the impact these support aspects will have on design and the importance of applying alternative design considerations early so that deployment costs can be minimised. Four specific diagrams have been highlighted. They are:

- Fig. 9.5 – cost effectiveness of diagnostics capability.
- Fig. 9.6 – corrective and preventive maintenance costs.
- Fig. 9.7 – maintenance policy costs.
- Fig 9.8 – total replacements in the life cycle.

The areas that are shaded show the costs directly related to items like repair, preventive maintenance or failures due to production or field losses. The curves present the approaches that should be used to measure the total cost of maintenance and losses versus repairs. The crossing points are the optimum approach in which the dollar values will be minimised from the total. It is important to recognise the need to plot these various curves and to allow the life cycle costing analysis to help the user access and determine the best design approach that is feasible for the product that is being produced.

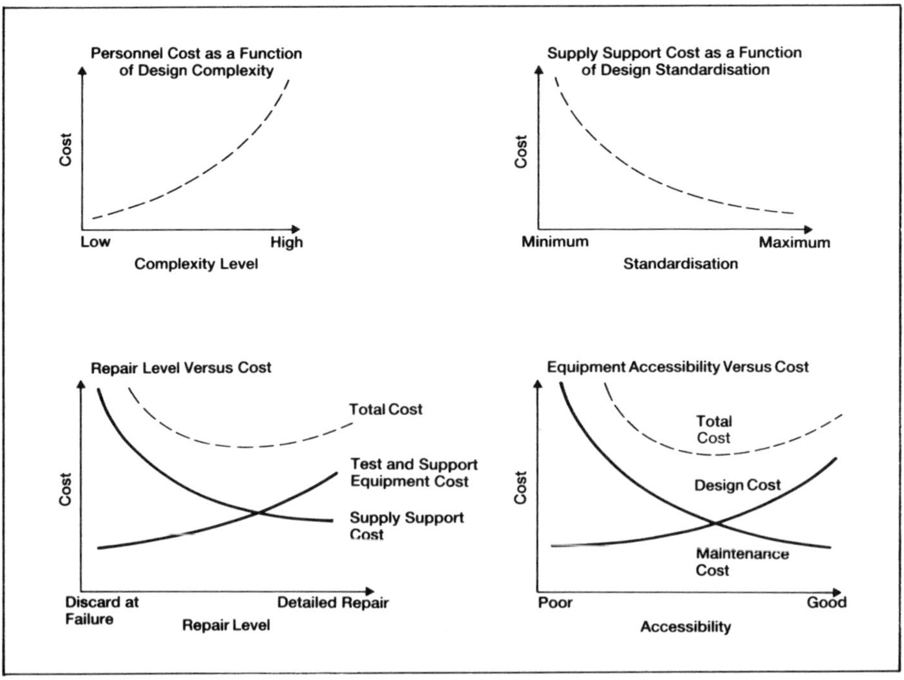

Fig. 9.4 System design versus life cycle cost considerations.

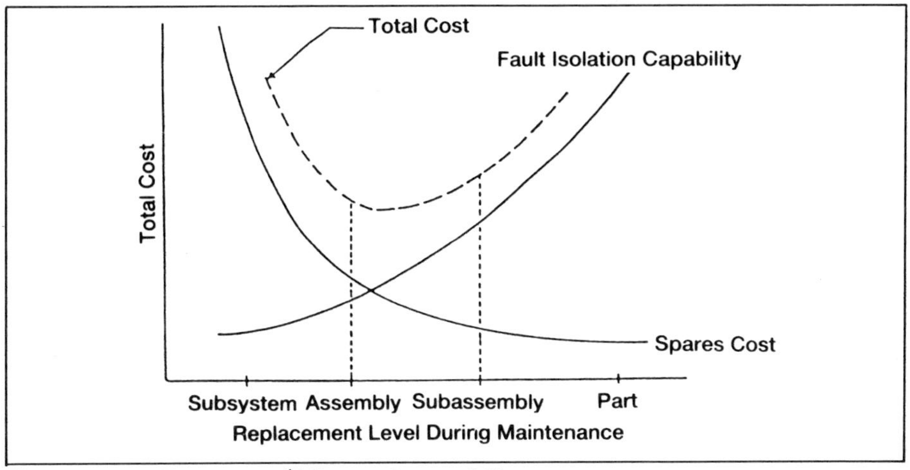

Fig. 9.5 Cost-effectiveness of diagnostics capability.

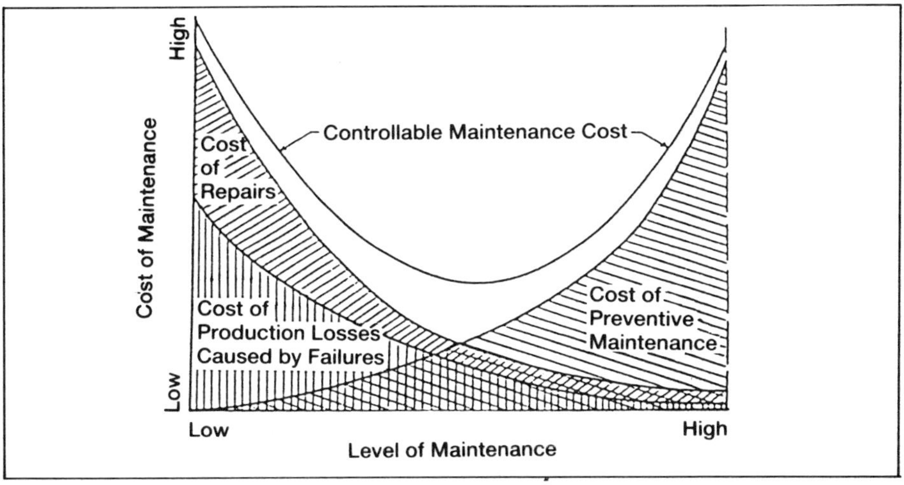

Fig. 9.6 Corrective and preventive maintenance costs.

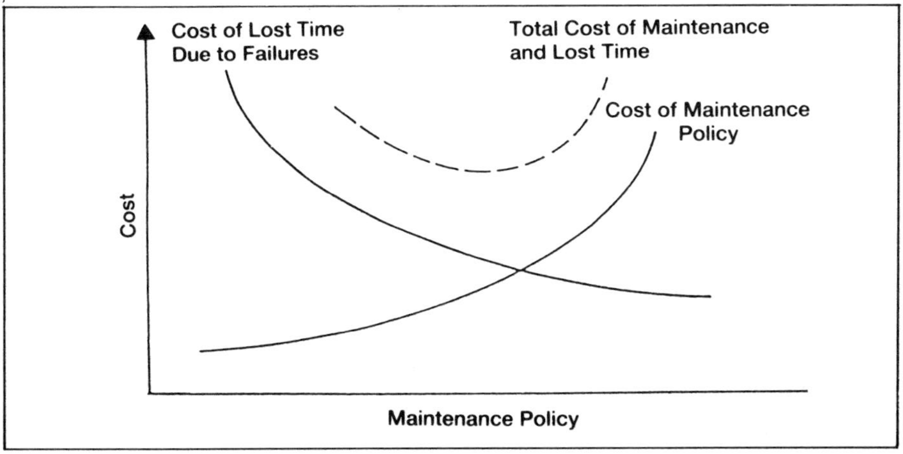

Fig. 9.7 Maintenance policy costs..

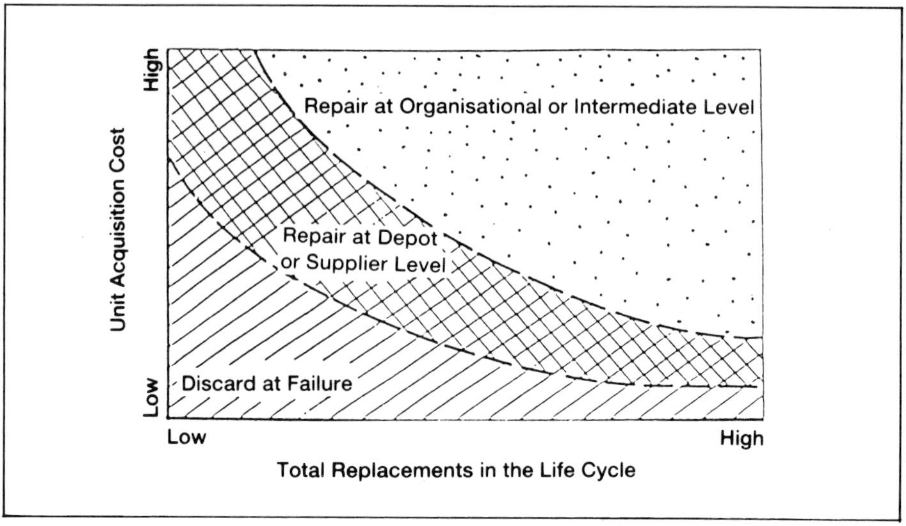

Fig. 9.8 Total replacements in the life cycle versus unit acquisition cost.

How can the point in the design be reached where these graphs can be plotted so that the direction the design approach is taking can be understood? That is the subject of the following paragraphs.

Cost estimating procedures

The LCC can be estimated using any number of estimating techniques. Cost estimation is refined as the programme matures. This is illustrated in Table 9.1. The source and type of the cost estimates usually depend upon the certainties of the cost elements. The following estimating techniques are commonly used:

- parametric analysis;
- analogy;
- bottom-up engineering;
- other (best guess delphi process)

Table 9.1 Cost estimating methods and most prevalent use by acquisition phase

	Concept exploration	*Demonstration/ validation*	*Full scale development Early*	*Late*	*Production*
Parametric	P	S	S	NA	NA
Analogy	S	P	S	NA	NA
Bottom-up/ engineering	NA	S	P	P	P

NA, not normally used; P, primary method(s); S, secondary methods.

On large programmes, most of these approaches can be used to estimate cost; however, bottom-up estimating is labour intensive and time consuming. Parametric analysis can be done very quickly, in a matter of days, and it correlates closely with estimates generally using the bottom-up method (usually within 2 to 3%). The analogy is generally a best guess that is not intended to be a guess for the total cost. Frequently, estimating the cost of a component with no precedent data is an educated guess.

The performance of an LCC analysis

LCC analysis is the study of LCC estimates in elements to identify life cycle drivers like the total cost to the government in procuring cost risk items and cost effective changes. It is an engineering tool with applications to all elements of the product. Cost modelling is often used to identify and analyse life cycle cost drivers. Cost drivers consist of control operational support costs such as reliability, maintainability, parts, and support equipment; and they are areas whose resources can be best applied to achieve the greatest benefit in reduced cost. For example, a reduction in manpower requirements would have a significant effect on a product's life cycle cost. An effective LCC analysis will identify areas where contract incentives may be applied to earn the largest payoff. Modelling for LCC is also useful in cost benefit and cost effectiveness studies. Long-range planning and budgeting, comparison of competing designs, decisions about replacement of ageing equipment, control of an on-going programme and selection among competing contracts are some of the ways to use LCC.

Many computer models are available, but no single model suits every application. This book does not identify specific models to be used; rather, it identifies the elements that should be evaluated to determine whether the LCC actually does what it is intended to do. The best approach is to use the model, experiments with it and determine whether it satisfies the specific requirements. If it does, it should be purchased.

Life cycle trade-offs

LCC trade-off analyses are normally employed to obtain what is considered an optimum balance between cost and effectiveness. A trade-off analysis that could be used is briefly discussed as part of the balance described in Chapter 2. Potential input variables for the LCC trade-off analysis are presented in Fig. 9.9 (for example, micro-processors could be incorporated into a design to automate functions and provide for built-in tests). This would

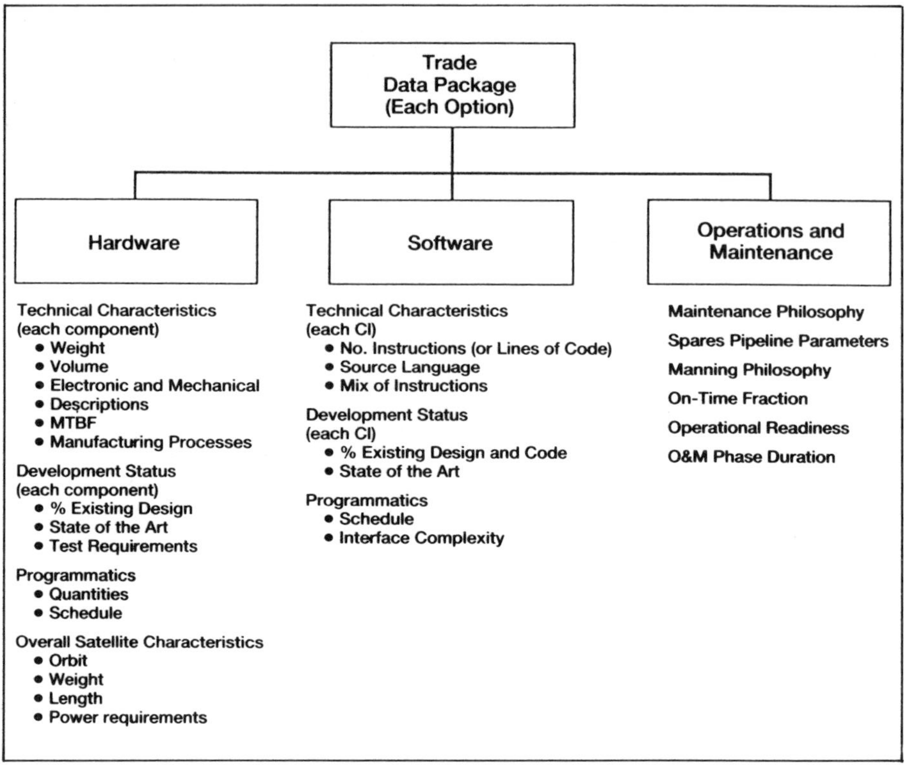

Fig. 9.9 Trade-off and input variables.

increase development costs (due to software development), but it could also reduce weight, size and maintenance costs with a potential over-all cost reduction. In considering LCC, reliability and maintainability have a major impact. The product and subordinate level unreliability reflects directly on the required logistics support in terms of skills and levels of maintenance personnel required, spare parts stockage, personnel training and the effects of operating on maintenance support. Applying this trade-off approach with the various inputs, a sensitivity analysis could be run to determine what areas produce the resources and how best to apply them to achieve the maximum cost benefit.

These studies examine performance parameters to determine where small changes in the parameters could produce significant changes in development and operation costs. As described in Figs. 9.4 – 9.8, various costs supplied to the elements described, if appropriately priced, could have a direct impact on the total support costs for the product. The intent is to ensure that the LCC model has an impact on logistics and logistics support. Therefore, the primary goal of the discussion is to ensure that the ILS programme is applied to LCC and that it can achieve a system readiness that is affordable. The resources needed to achieve the readiness objectives must receive equal emphasis with the resources required to achieve a schedule of performance objectives. LCC analysis helps to achieve these by evaluating the cost implications of the various design and logistics support alternatives. One cannot stress too often that the LCC should be applied early in the design cycle and concentrate on qualifying the cost implications of the design alternatives that provide the desired level of performance. ILS activities at this stage focus on designing supportability into the system and evaluating the cost of ownership and support requirements.

As the design progresses, evaluations are then refined, identifying lower cost means of support to achieve readiness objectives. In particular, support elements such as manpower and spares are evaluated to identify cost effective alternatives by which required readiness levels can be achieved and sustained during actual operations.

Time phasing of LCC policies

Cost analysis in the early design phase is typically concerned with making initial estimates of total system costs, including the alternatives to new product acquisition, modification of existing or use of commercial products, and changes of policies or doctrines. The object here is to estimate all elements of LCC for use and comparison of alternatives. Therefore, these cost estimates will reflect analysis of pertinent supportability factors for the alternatives proposed, with adequate attention given to costs associated with the risk of uncertainties of each alternative. Results of these analyses should show the order of magnitude of the costs of the alternatives in comparison with product modifications and the use of other existing equipment.

Cost estimates based on detailed design engineering generally are not necessary at this time and may not be either possible or available. By the time the actual design has been determined, the logistics personnel must develop specific plans to analyse support costs and readiness drivers of current fielded systems and readiness and support cost targets for improvements based upon the analysis. Careful planning must be directed to the development and analysis of the data for the current fielded baseline equipment. In some cases, an existing database can provide the information, though it may have been developed through other means for other products. Generally, the cost analyst must recognise that the data drawn from other products may not be consistent and could require some adjustment before being utilised in the comparative study.

This earlier phase, the initiation phase, is when marketing, field service and other outside factors can be quickly analysed and measured to determine how best to improve an existing product or field a new one that will satisfy the consumer's needs. Once this has been decided, the next phase in our design is the conceptual phase. This is the point where we start designing a piece of equipment to determine how the product will function. We are only looking at the functional aspects of the design to ensure that the technology is available to produce the results we are looking for. For discussion purposes, we will call this the concept phase.

The LCC objectives in this phase concern themselves with developing cost estimates for each alternative concept, demonstrating affordability and identifying cost drivers.

It is now time to integrate LCC into the LSA approach and to take a look at those tasks and determine which will have a direct impact on the LCC itself. The logistics manager or the cost analyst should participate in structuring and tailoring the LSA and LSAR requirements to obtain sufficent data for later LCC analysis. Explicit plans must be written so that trade-off studies are performed that are designed to set firm goals and thresholds for selected parameters. During this design phase, the programme accesses the supporting cost analysis and seeks to identify areas requiring additional design action because of unacceptable estimated cost levels.

Through this entire acquisition cycle, cost reduction alternatives derived from value engineering, producibility engineering or alternative operations and maintenance concepts require the increased use of commercial equipment, and industrial modelisation incentives are considered to keep costs at or below stated goals. These must be incorporated in the cost analysis programme to provide the engineers and the logisticians with the cost implications of their alternative design and support concept. Cost reduction alternatives must be considered as early as possible in this acquisition phase because this offers the greatest opportunity to reduce the total life cycle cost of the product.

Proceeding to the design and development phase, other risks become apparent. These need to be analysed and cost analysis tests performed (including trade-off studies) to ensure that all the approaches during this design activity are taken into consideration. The cost analysis performed must provide credible estimates of the relationships of acquisitions versus support costs, the changes in logistics support and design alternatives. The results of these cost trade-off analyses are most useful when the estimate provides sensitivity data that includes the cost range or exposes the cost risk areas associated with engineering and the support alternatives (see Figs. 9.4–9.8).

As the design proceeds, the parameters for the study become more defined and reliability becomes a more pressing factor in reducing operational and support costs. Since the ultimate logistics objective is to achieve a system for economical cost, the reliability that is sought by design in the range of minimum life cycle cost. There is an economic and a technical limit in designing reliability improvements. Theoretically, the programme development strategy will result in the balance of acquisition and support that produces minimum life cycle cost (see Fig. 2.5).

Full-scale development

By this stage, sufficient logistics support analysis data are becoming available through support cost analysis at the subsystem and even the component level. LSAR data are particularly helpful in estimating two of the largest supporting costs: spares and manpower. The LSA records contain estimates of maintenance man-hours, repair parts consumption rate and requirements for support equipment, training devices and facilities. Management of the LCC also requires that explicit plans be developed for cost analysis updates during the subsequent operations. Fig. 9.10 shows the connection between the LSA and life cycle costing. The central portion of the diagram, defined as the database management system (DBMS), establishes the mechanisms by which data can be moved between and among the various files available to support the total life cycle approach. Therefore, the tie between LSA and the LCC becomes extremely important, and it is necessary to ensure that the data generated by the elements located on the left side of the diagram (engineering and support management) are properly controlled and upgraded so that an accurate LCC file can be maintained and an LCC output document can be produced.

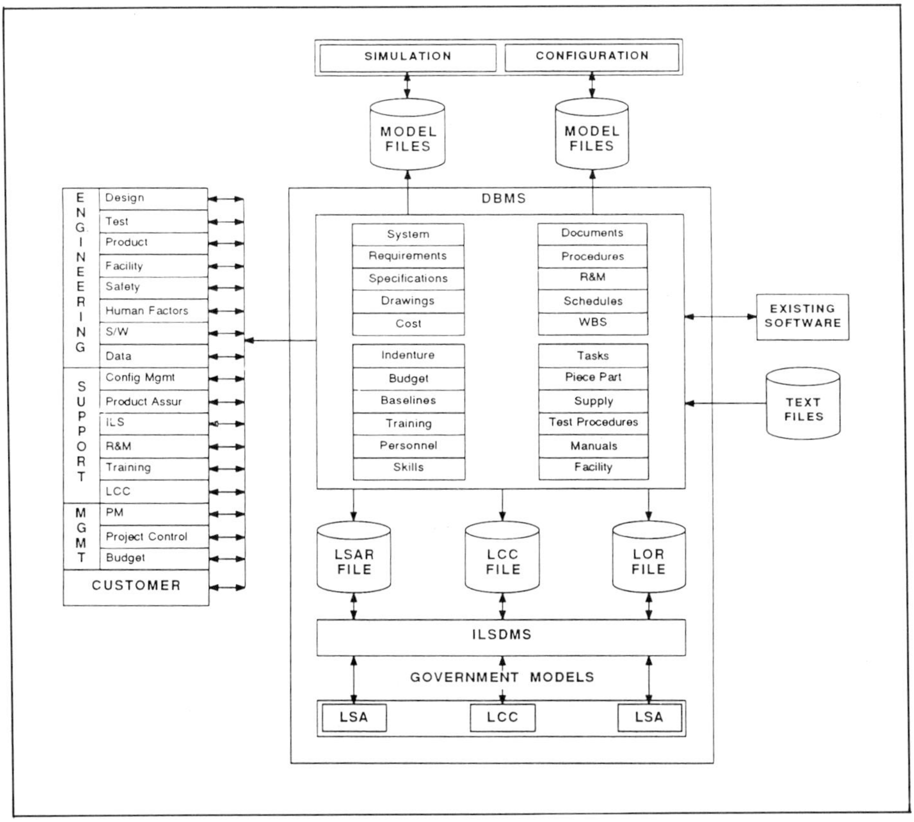

Fig. 9.10 Logistics support analysis/life cycle cost approach: an overview.

During this phase of the programme, discrete estimates can be performed on the individual pieces of equipment being designed for the product. One must be aware of the risks uncovered during the previous phase by quantifying their potential cost and ultimately demonstrating their affordable resolution. The cost estimates during this portion of the acquisition should be more accurate because they deal with more detailed issues and, in particular, become a significant factor in the production discussion. By the completion of this phase, an accurate total cost should be produced and made available to the management staff so that they have an appreciation of what to expect over the long term.

Summary

The lack of LCC will have a direct impact on design and logistics support processes. To be most effective, the integrated-into-the-engineering process that amalgamates the design and logistics choices must start with a programme initiation. Once the ability to influence the design is lost, it is very difficult and much more costly to re-establish. Most performance and schedule risks involve costs. Performance risks result from requirements that are very costly or from engineering requirements beyond the foreseeable capabilities for both hardware and software development. The result can be increased costs for the design, development and test of a replaceable item, contract termination, programme buys; and increased support costs.

The importance of ensuring that the programme manager encourages the designers, logisticians and other cost-applicable functions to work together cannot be overstressed. Early life cycle cost analyses should be performed so that the programme manager will be able to determine the impact that the design changes are having on his product. Life cycle cost exerts direct influence on design. It also requires engineering participation in logistics analysis inputs. This will directly affect readiness and the costs that influence operation and support.

The implementation of LCC concepts is a way of doing business in today's environment. This implies that management design and all of the other supporting elements using computerised approaches must be used. These new products are so complex and support costs so high that attention must be paid to all of the detailed design features that enhance product support, increase operational availability, and lower life cycle cost. At the same time, this must be accomplished at a reasonable development cost.

Chapter 10 VALIDATION AND VERIFICATION OF THE LOGISTICS SUPPORT SYSTEM

The purpose of an ILS validation and verification programme is to provide assurance and proof to the consumer that the support system development can provide the required support to a product in its operational environment. The basic objective is to ensure supportability under anticipated environmental conditions. The second is to verify that the logistics support developed for the product is capable of achieving its established readiness levels. The final objective is to demonstrate that the readiness objectives are obtained in all kinds of conditions, regardless of the product's utilisation rate during operational use, and to prove that these goals have been achieved when testing is started.

That is also a very basic question to the logistics engineer, as it is to the design engineer, because the testing should start as soon in the design process as possible. The testing does not have to take on the aspects of actually fielding a piece of equipment and checking it out in its environment, but can consist of testing through the use of models and analyses. These two essential tools are used to ensure that the process that is being applied to support testing or design has been thoroughly analysed and simulated wherever possible to ensure that all aspects of the design have been properly treated and traded off.

The modelling and analysis will give the logistician the ability to determine how well his alternative approaches to the design and logistics support are being welded into the overall product. As the design progresses, the testing aspects may be validated through the use of informal bread boards, brass boards or prototype development models or through the use of models, modelling and analyses. It is continued until the design has been completed. Besides the informal bread boards and brass boards, formal tests and demonstrations are accomplished during the latter part of the development phase prior to preproduction. It is normally done on prototype equipment, which will be similar to the production equipment but not necessarily fully qualified at that time. These tests are generally conducted by the contractor or manufacturer and witnessed by the customer.

Operational support equipment and preliminary technical manuals are used for test support. Specific types of tests include formal maintainability demonstrations, support equipment compatibility tests, personnel tests and evaluations, and technical manual verification and validation. Test data are analysed to determine whether the equipment configuration can be changed to eliminate maintenance requirements' problems and whether the support system will in fact satisfy the maintenance definitions. The last are the formal tests and demonstrations prior to large-scale production. These are conducted by the customer's personnel at his sites. Contractor or manufacturing personnel provide certain predetermined on-site support. Operational support equipment and spares, and formal technical manuals are used. Field test data are collected and analysed to determine whether the product meets all maintainability and maintenance quantitative requirements. This is the first time that all elements are operated and evaluated as a support system, and is an opportunity to access total product design from a support standpoint, as well as to access the product in terms of the specific support requirements (quick turnaround times, pipeline times and so on).

The last aspect is the total operational test and its associated support systems conducted in a true field and maintenance environment. These tests involve customer personnel, actual live facilities, operational support equipments, spares and repair parts and the appropriate final technical manuals.

The tests just described answer the question of how logistics are verified. It was initially stated that the first portion of verification logistics occurs through analysis. An attempt is now made to apply the safety and human engineering aspects, and the other supporting engineering functions to the logistics system to determine:

- whether the maintenance concept has been appropriately defined and is working in the role in which it was originally designed;
- whether all of the functional aspects of the equipment have been properly organised and placed so that the operator can easily read and determine what he is supposed to be seeing. The review and analysis of the

data being drawn (for example, front panel readout systems, gauges, valves and so on) must be readily accessible to the operator and the maintenance personnel for appropriate use.

The demonstration process

Tests are also done through the demonstration process. The reliability and maintainability demonstrations are perhaps the most stringent tests performed in the static mode. The reliability test is performed under theoretical environmental conditions. Depending upon the size of the equipment in question, the equipment itself could be placed in an environmental chamber where it is baked, humidified, temperature cycled in an appropriate set of sequences and cycles to determine whether the equipment will satisfy the requirements stipulated by the specification. These tests will help prove that in the contractor's facility under semi-static or dynamic mode the reliability of the product can in fact achieve the requirements established for the programme.

If these tests prove otherwise, redesign or reconfiguration of the product will be required to satisfy the customer's requirements. The maintainability demonstration is similar in nature. The attempt here is to prove that the maintenance technician, using the training afforded to him by the contracting facility, can actually repair a fault or a simulated fault under standard operating conditions. To prove this, a series of malfunctions are installed into the product itself. These are done in series, one at a time. The intention is to show that the diagnostics, along with the training and technical manuals, are of sufficient depth and breadth to allow the maintenance technician to repair the product properly and bring it back into operational use. What the maintainability demonstration attempts to prove is that the manufacturer has been successful in designing the product to achieve the quantitative requirements established as part of the specification. He in essence is proving that the mean time to repair has been achieved. If this test, like that of reliability, does not successfully meet the requirements, the manufacturer is responsible for determining the reasons why the MTTR has not been achieved and takes the appropriate corrective actions to ensure that, for whatever reason, the equipment, personnel training or supporting technical documentation is improved so that the maintenance technician can rectify whatever problems may exist in the product within the mean time established for the contract.

These two tests are normally performed prior to going into full production. They are part of the contractor's test and demonstration phase and are carried out under full-scale engineering development models. The last set of tests on the full-scale engineering development model are described in a series of full formal tests and demonstrations prior to the production commitment. They will be performed in two situations. First, testing is conducted in the contractor's facility under the semi-static modes or semi-dynamic modes that lead to the acceptance of the product by the customer prior to his testing it in field conditions. These tests will fully exercise the product using all of the available signals necessary to demonstrate that it is operating to the fullest extent to meet its intended use.

Once this has been proven by the contractor, the customer then receives the product and places it in its intended environment, testing it on the truest dynamic mode possible. He is attempting, from a logistics point of view, to verify that maintenance is appropriate for the design and that it satisfies the approaches necessary to ensure achievement of mission requirements. He is also verifying that the training given to the maintenance and operational personnel was of sufficient depth to provide appropriate operations and support once the product is outside the contractor's domain. This will ensure that follow-on production training has been and will be appropriately updated as required to ensure that the training techniques found deficient during the actual dynamic testing have been corrected. This in turn ensures that the training courses are finalised and that the personnel required to teach the follow-on operators have the most accurate data available.

It also verifies the technical manuals in the dynamic operational test. The operation and maintenance personnel who will be using the technical manuals to follow-up on the operations that they are required to perform will assist in determining whether the manuals have been accurately prepared and that the documents accurately represent the configuration of the hardware. Any variations in the technical manuals should be corrected prior to final completion and deployment with the production units. Lastly, it verifies the test equipment, if this is available during the preproduction testing. If it is not, then appropriate standard test equipment is used. It will assist and dynamically check the internal circuits of the design, ensuring that the test points have been properly located and are easily accessible. These tests not only check the application of the test points and the functional diagnostic routines, but they also verify that the appropriate test equipment has been chosen to do the necessary jobs. Any variances in any of these specific areas could cause the product to be rejected and changes to be instituted prior to going into production.

In an attempt to preclude any unforeseen problems, it is strongly recommended that test and evaluation plans, policies and procedures be written, so that when tests are performed there will be a complete understanding by all parties involved as to what is intended. These plans, procedures and guidelines will establish the cycle in which the testing will be performed, measure its impact and establish the types of reporting systems required to identify both the positive and negative aspects of the tests themselves.

The logistician should participate in every way possible as the tests are being performed so that when possible problems within the logistics support system arise, he will have an appreciation of what may have caused this to occur and how best to correct them. It is normally his job to ensure that prior to going into testing, all of the appropriate equipment, spare parts, tools, personnel, technical manuals, and so on are made available so that the operating tests can be performed without having any undue delays. He will be measuring the human impact of the design along with the engineers to ensure that the maintenance personnel can actually perform the necessary corrective maintenance tasks to keep the equipment operational.

He should stress to all of the management personnel involved in the test that he is there to assist. The data that are derived from the tests will be flowed back into the logistics support analysis database and measured against the predicted modes of suggested approaches, and the variance between what occurred in testing and that predicted as part of the LSA will be determined. Any variations between this and the LSA process and the actual testing will be scrutinised by an evaluation team to determine how best to implement the appropriate corrective action.

It is sometimes very difficult to understand the various levels of tests that are performed for each phase of the development of a programme. It is easy to narrate or describe them, but it is a lot easier to understand it in pictorial form. Fig. 10.1 is a modification of Benjamin Blanchard's description of the various phases of the testing cycle, describing testing for each of the phases.

In all the preceding discussions, we used testing in a more global base than would normally be expected or anticipated. Testing is a design development approach, and is performed on a product baseline, but one must recognise that testing does occur early on, and that certain aspects of the

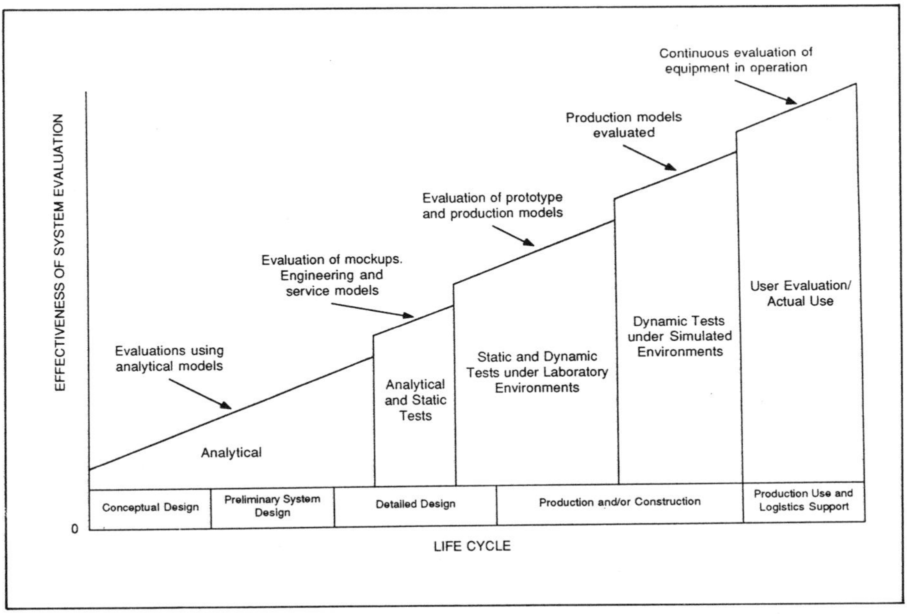

Fig. 10.1 Testing during the life cycle.

cycle need to be done independent of the actual integration of all of the parts. One aspect is the development and testing of software, which has become one of the most important pieces of our more complex designs.

Software development process with the emphasis on testing

Software has become one of the major elements in the development of most products in today's market. Whether it be in a single chip with firmware or a complete computer capability built into the product to support a majority of the operating functions, software has become and will continue to become one of the most dynamic aspects of all of the designs off the drawing boards. Therefore, the logistician should have a good appreciation of what the software development process is and when software can and should be tested, and how best to be able to maintain the configuration item product base. This portion of the logistics profile can be a problem because any minor change in one piece of software or in the coding itself could have a direct impact on all of the functional aspects of the system. Therefore, the logistician should not only have a good appreciation of how the hardware works, but also of the various pieces of the software that have a direct effect on his efforts. Specifically, in text books associated with the operations of the product, detailed information in the technical manuals for operational procedures should describe some of the software techniques being utilised to assist in the operation.

Before discussing this, let us go back and emphasise that there are certain pieces of the puzzle that should be looked at. The common procedure in putting a jigsaw puzzle together is to define what the frame looks like. In essence this is done in the preparation for software development. The framework in this case is a description of the general documentation of software requirements. This is no different because it is a result of the design requirements for the total package being developed. Out of the software requirements the preliminary design and detailed design lead to the document called the Software Development Notebook. This essential book will become the basis from which all future documentation is generated.

Fig. 10.2 describes the general flow of the work being performed. The centre portions of the software support packages are the test plans and test procedures. Various testing is performed on software well in advance of any elements being tested within the hardware structure, because testing of software elements can be easily done. It is similar to testing a function within the design on a brass board, which could be equally defined as testing a series of algorithms within a software program computer model. This is done for the same reasons as in the design of a piece of mechanical electrical hardware: in order to establish the building blocks, the documentation from which all coding will be generated.

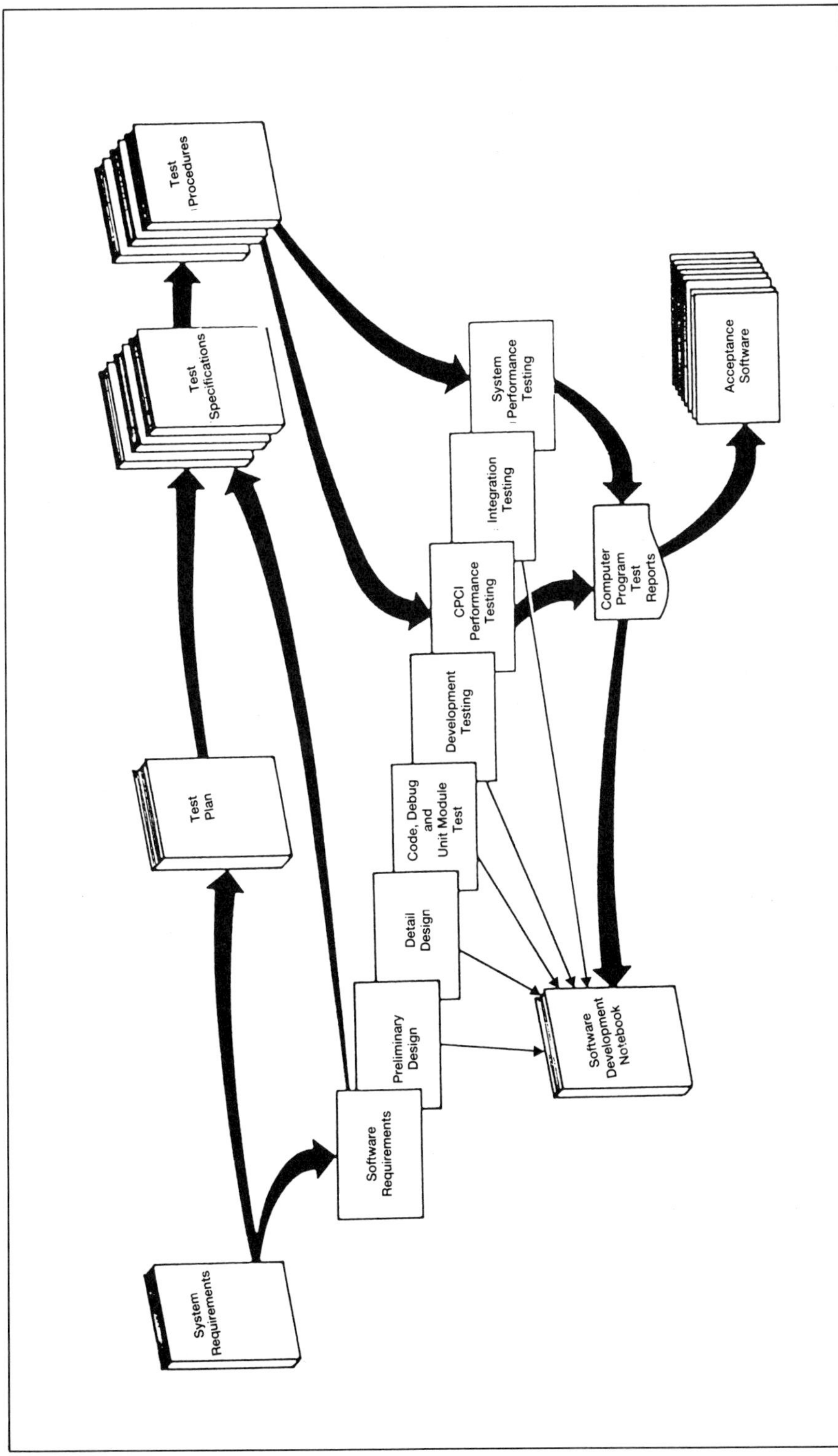

Fig. 10.2 Software development process with emphasis on testing.

No coding should be started until all the building blocks or the foundation, be it the software requirements specification, the preliminary design specification or detailed design documentation are ready. Test planning of software should take a very quick and early role in the development of the entire programme. The generation of appropriate techniques to verify and validate that the software is being generated and created in a qualitative manner is essential. Because of complexity, the reliability and maintainability of most of the equipment that we are now distributing and using are becoming very uncertain. Some of this is due to the inability of logisticians and quality assurance to accurately check and validate the fact that the software has been appropriately debugged and that the necessary corrective action has been incorporated into the design. Therefore, the purpose of testing is to ensure that as many types of inaccuracies or errors as possible are detected, including incorrect requirements design or coding not detected by previous verification activities.

There are two aspects of the validation and verification related to testing. The first is the independent planning and analysis in reporting the test itself: the essence of the traditional validation and verification function. The object here is to exercise the various capabilities of the evolving software system. Second, and no less important, is the necessary monitoring and review of the progress of the testing performed. The validation and verification staff (comprising the logistics, software, systems and design engineers) should determine how well a test is progressing, how thoroughly the evolving software is being exercised, whether the test results are being properly interpreted and whether these test results are conclusive. Lastly, they must determine that the test reports are thorough and accurately reflect the test. Above all it must be remembered that testing is an inherently difficult activity, since its essential purpose is to discover errors in the programme. The validation and verification functions pertaining to software are associated with the following:

- The verification that the system meets all of the stated requirements and specifications. This holds true for the system level but for this discussion we refer only to software systems.
- That the application and use of the appropriate software is demonstrated, and that it meets the user's interface requirements as he expected, validates inputs to produce and produces the expected output. Any illegal outputs should be handled in a prescribed manner without causing the system to fail.
- Production of sufficient feedback at each stage of the development so that the major design changes will not be necessary in the implementation of the final stages.
- Verification of modules, databases and interfaces during all stages of system integration.
- Validation of test drivers.

- The preparation of all aspects of acceptance testing including scripts of scenarios, report documentation of selection and preparation of supporting test equipment.

Test and evaluation has always been part of the validation and verification of the software program. In this area, perhaps more than any other, the necessity for an independent V&V group should become evident. In designing or coding most programmers generally have enough professional pride to subject their work to close critical analysis to discover errors or inefficiencies. When it comes to testing, however, there is a strong tendency to run 'quick and dirty' tests and get on to the next program. Hence, the necessity for an independent test group cannot be overemphasised. If this is not done, the logistician will have an immense problem on his hands once the equipment has been deployed and the product starts experiencing failures. These failures will not be able to be quickly traced to whether the problem lies within the hardware or the software. In addition, the interfaces among integrating functions with other vendors become an increasingly delicate point: whose problem is it? How do you validate and verify that one soft-ware problem isn't the other person's hardware problem, or has been caused by his inappropriate approaches for testing in the product designed? Therefore, the logistician and all the people he works with should ensure that the various tests that are necessary are constantly performed.

There are different levels and types of testing to be performed during the generation of computer software. They are as follows:

- *Unit module testing.* This is normally used for debugging of individual units in isolation. In other words it is operated individually, to make sure it works before integrating it with other functions or operations of the equipment.
- *Self program testing.* The integration of individual modules into subsystems demonstrating major functional performances within the CPCI (computer program configuration item).
- *CPCI performance testing.* This is a validation of the entire CPCI against the requirements as specified in the program performance specification. This test is a demonstration of readiness for integration.
- *System integration testing.* This is a verification of product integration of software programs with the equipment and with other systems, as well as man–machine interfaces in the demonstration of functional completeness and satisfactory performance.
- *Product performance software quality testing.* This is the comprehensive testing of an entire system and its intended operational environment against the systems specification. This test demonstrates readiness.

The work just described as part of software development is no different from that performed for the development of hardware. What is unique, though, is how to integrate the hardware with the software or the software with the hardware, so that the integration procedure is done properly. Any wrinkles in the design will have a direct impact on the ability to integrate the hardware and software properly together. The question really comes down to what gets designed first, the hardware or the software. This is like the chicken and egg syndrome. It depends on where the emphasis is placed, what kind of company you're working in, and how appropriate it is to try to integrate both functions. This is determined on the effectiveness and the efficiency of the management involved. One must recognise early on that no matter how much testing is carried out, either independently or dependently, there will be problems. Operational use of the software often uncovers errors that were not discovered prior to accepting the equipment. Part of the reason for the situation is that it is virtually impossible to test completely all possible operational events prior to deployment. Some errors also occur due to inadequate or inaccurate documentation, so that the software does not operate as specified in the documentation. In any case, when an error occurs, a software trouble report should be made and filed. This trouble report is the same as the trouble or failure reports made during all of the other phases. It becomes an integral part of the total design and is used to evaluate the acceptability of the product to be used. As the failures accumulate, it will be easier to go back and retest the design to ensure that there are no latent defects, and that as the changes are implemented appropriate tests are performed and software modules are integrated into the system, again being rechecked and tested to ensure that the integration is successfully accomplished. This cannot be overstressed.

Backup during testing is essential. Failure in the software program could cause the design to go down, resulting in complete loss of all of the data that have been previously collected. Therefore, prior to the integration of any new test modules, one must ensure that a backup program has been prepared and is easily accessible. There is a dramatic change in the environment when attempting to rethink the system.

Summary

Test and evaluation therefore are an absolute necessity to access the acquisition risks and to evaluate the operational effectiveness and suitability of the equipment being fielded. They also constitute a non-destructive evaluation to determine how well the logistics support, and the reliability and maintainability requirements, have been achieved. The non-destructive logistics testing is performed at the manufacturer's facility and then proved during the field execution of the actual test performed by the consumer. All of the tests are necessary for total acceptance of the product that will eventually be used.

The purpose of the test and evaluation is to prove the validity of the actions, analysis and assessments undertaken during the earlier part of this specific programme. Test and evaluation are used to demonstrate operational suitability, being able to state that the logistics that were defined, analysed and integrated in the system actually support the specified baselines. The testing from a logistics point of view validates the need for all the requirements stipulated in the specification, and proves that the various alternatives that culminated in the actual design satisfied the programme. It minimised the cost and maximised the benefits so that all of the logistics support requirements were accomplished. It presented to the consumer a realistic and timely approach by which all of the elements were fulfilled. This is the area where the 'rubber meets the road'.

Chapter 11 USER SUPPORT

When you bought your last car, do you recall the sense of relief you felt when the salesman assured you of the wonderful service you could expect when you needed it? Do you also recall the first time you took that new car in for minor adjustments and service, only to become disenchanted with the treatment you received, the poor quality of work performed and the unexpected out of pocket expenses for items 'unfortunately not under warranty'? Disgusted, you left the agency with the thought, 'What ever happened to user support?' From then on you take your car to a local service station.

Contrast this situation with the theme of William Twiner's popular book, *Secrets of Personal Persuasion* (1985). His theme stresses that if you do not penetrate the barrier between yourself and the customer (audience), the barrier will grow larger. 'Get off to a sloppy start, and recovery is difficult, credibility is submerged.'[1] This sound advice certainly applies to industry and service. User support, though not high on the priority lists of most manufacturers, is mandatory to retain a competitive advantage. Product and service development does not end when the product or service is provided to the customer. It should last throughout the product or service life cycle. The professional logistician recognises this and seeks to ensure that user support is provided and considered from the conceptual design onward. User support requires a sense of accountability on the part of companies that market their products and services.

[1] Twiner, William (1985) *Secrets of Personal Persuasion*, p. 7.

Customer services

Pricing

Customer service is not usually associated with logistics. This aberration questions the route from raw facts to official results in terms of quality, excellence and service. Cost and profit generally prevail with retailers, manufacturers and in service industries. Prices are set at levels to cover costs and maintain operations, and range from too high (good margin, inadequate volume) to too low (good volume, inadequate margin). Thus, pricing depends on consumer-perceived value based on what customers will pay and what quantities are demanded.

The price affects sales revenue and profit. It should take into account two major elements:

- the specific marketing programme;
- consumers' desires as to price value.

These two factors have the following components:

- *Specific marketing programme product*
  Product performance capabilities
  Appearance
  Convenience
  Variety
  Designed-in obsolescence
  Quality in excellence

  Service
  Installation and assembly
  User application assurance
  Return policy
  Warranty and guarantees
  Logistics support

  Purchase situation
  Product availability
  Quantity
  Credit and instalment paying arrangements

  Physical distribution and transportation
  PD&T strategy
  Warehouse management
  Transportation carrier selection and costs
  Distance and time requirements

Image
Brand recognition
Brand image
Reputation for excellence and quality
Reputation for customer service

- *Consumers' desires*
  Premium, standard, discount, and leasing price product and distinctive service characteristics.

A special comment should be made about discounting pricing and leasing. Discount pricing is based on low costs, high volume and lower prices. It consists of the following:

- Reduced labour prices.
- Lower product costs.
- Less expensive facilities and equipment to merchandise.
- Minimal promotional expenses.
- Fewer customer services.

Leasing, on the other hand, is increasingly popular. Leasing consists of the following:

- Minimised large capital outlays, which offset the losses of capital depreciation vastly reduced under the US Tax Reform Act of 1986.
- Reduced labour costs.
- Reduced facilities and equipment expenditures.
- Some guard against the effects of inflation.
- Capability to modernise without costly expenses.
- Sound contracting and acquisition management required.

Overdoing cuts in user service

The high costs of running facilities, labour and ineffective managment have brought about the trend to cut user services. But manufacturers, retailers and service industry providers can go too far in cutting cost and reducing services. Treating the customer as a VIP costs little more, and providing them with some recognition of the value of their business normally creates more volume since business is increased by word-of-mouth.

With the decline of middle-class purchasing and increasing older segments of the population who also have declined economic power, an unusually large segment of firms has expanded into the well-to-do, lower income and youthful credit-card-user markets. These firms believe that competition is less severe at the more wealthy and youthful ends of the market. Firms are promoting brand name labels and price lines. This recent trend poses unique problems for the logistician.

The logistician's challenges

The logistician in the marketing programme described above is faced with the following challenges in his or her job:[2]

- Maintain quality and excellence in the product or services for minimal, optimal cost.
- Ensure that maintainability and substainability of the product or services are effective and meet the requirements of the user.
- Forecast the required spares and consumables, to include corrective and preventative maintenance as derived from the logistic support analysis (LSA).
- Determine corrective maintenance costs, to include related spares, test and support equipment, transportation training, facility costs and necessary travel budget.
- Calculate costs associated with preventive or scheduled maintenance.
- Determine transportation and handling costs, to include return of faulty items, overhaul and calibration if necessary.
- Allocate initial technical data preparation costs.
- Determine product or service system and equipment modification costs throughout the system life cycle to improve performance, effectiveness or a combination of the two.
- Determine product or services' reliability, to include defining system and equipment wear-out periods, identify failure modes and failure rates, determining the product's mean life, and protection against secondary failures.
- Determine product and service system phase-out and disposal costs, to include the liabilities or assets that occur when the system is terminated and disposed.

Managing user support

User support requires logistics support management that includes planning, organising, directing and controlling all of the activities associated with production of the product or services and use of the product and service by the consumer. This management responsibility begins as early as the conceptual design of the product or service and continues throughout its life cycle. The system life cycle management will differ based on the nature of the product or service, but for most there are six basic programme phases:[3]

---

[2] Blanchard, B. (1986) *Logistics Engineering and Management*, 3rd edn (Englewood Cliffs, NJ: Prentice Hall).

[3] Ibid., pp. 5–9.

1. *Conceptual phase.* This is where the product or service system is defined, acquisition requirements are determined, production and operational support requirements are identified and logistics support planning is established.
2. *Advance development phase.* This phase identifies and defines the product configuration. It deals with product objectives, performance limits, areas of risk, alterations and acquisition methodologies. During this phase, logistics support analysis is conducted to verify design concepts, followed by the initiation of the logistics support plan.
3. *Detailed design and development.* This includes the design and test of the product. This phase consists of design effects on reliability, maintainability, sustainability, human factors issues and logistics support. Engineering modes and prototypes are developed. At this time, the formal logistic support plan is completed.
4. *Production or construction.* This consists of producing the product and its support, such as test and support equipment, spare parts, training personnel, software, physcial distribution and transportation, and warehousing.
5. *Operational use.* This covers all of the activities associated with consumer use and logistics support, to include maintenance and sustainability.
6. *System retirement.* This consists of all of the actions necessary to phase out and dispose of the system, to include waste management, recycling or other responsive actions.

Logistics planning

To effectively manage products or services development and use by customers, good logistics planning is required. Among the functions to be covered during this planning process are:

- An evaluation of design and logistics support in terms of product requirements by conducting feasibility studies and maintenance and support policies.
- Development of functions and tasks according to the product and its requirements.
- Task scheduling, to include levels of activity and milestones for completion.
- Development of activity networks in accordance with the tasks to facilitate management and control. (Use of PERT charts facilitates this process.)
- Identification of funding sources and preparation of cost estimation.
- Development of the organisational function to accomplish the tasks identified for the project.

Throughout all of these functions, user support requirements must remain at the forefront of the planning and management process. The end result is the logistics support analysis (LSA) and the formal logistics support plan (LSP). Both are ready for use beginning with the design and development phase and subsequent phases.

The logistics support plan

The LSP is fully developed from the preliminary logistics support plan (PLSP), and covers all of the logistics support activities throughout the product's or service's system life cycle. It covers the following as a minimum:[4]

- *Detailed maintenance and sustainability plan* – as developed from the LSA information.
- *Reliability and maintainability plan* – includes analysis and prediction information.
- *Test and support plan* – covers considerations for acquisition; make or buy issues, testing and evaluation and a delivery schedule.
- *Supply support plan* – includes pre-operational and operational support, also LSA data on maintenance, spare parts, warehousing and materials handling.
- *Physical distribution and transportation plan* – covers packaging, storage, transportation and safety criteria.
- *Technical data plan* – to include technical data needs, technical development schedule, verification and validation plan, methods to handle changes and change documentation.
- *Facilities plan* – lists all of the real property and equipment necessary to support the system testing, training, operations and logistics areas.
- *Personnel and training plan* – covers personnel requirements for operators and maintenance.
- *System retirement plan* – covers phase-out of equipment, restoration of items and disposal of material in accordance with effective waste management procedures, EPA, OSHA and social responsibility.
- *Detailed management plan* – covers all the elements requiring management and control.

All ten of these subcomponents must include user support as a fundamental requirement for production and operational use.

For further development and treatment of the LSP, the reader's attention is directed to Benjamin S. Blanchard's *Logistics Engineering and Management* (3rd edn, 1986).

Summary

User support is critical to the success of the firm. Every company reflects its excellence and quality through consumer perception of its products and

[4] Ibid., pp. 447–461.

services and the way it stands by them. The effective management of the producer/consumer relationship should be more important than short-term profits, as is the trend today. The application of logistics principles will promote better success for the firm if it keeps user support in mind. The key to this success is through effective pricing, not overdoing cuts in user services, meeting logistics challenges, managing user support and logistics planning.

The day of abundant resources, beneficial economic conditions, production arrogance, bad management and lack of accountability to the market and consumers is over. Firms planning to make it into the twenty-first century are going to have to change their *modus operandi*. They can start by utilising logistics support that focuses on the consumer.

Chapter 12 OBSOLESCENCE PLANNING AND WASTE MANAGEMENT

The most ignored dimension of corporate management is the failure to conduct obsolescence planning and waste management. Facilities, products and services all become obsolete and must be disposed of. Disposal and managing waste is largely unattractive to most people. The throw-away disposable mentality of most of us is demonstrated by overloaded dumps, garbage disposal problems, excess waste of material, new concerns over the depleting ozone layer which is producing a 'greenhouse effect' all over the world and an overabundance of industrial and vehicular pollution everywhere.

Obsolescence occurs because of manufacturer's design or because of worn-out equipment. Designed-in obsolescence is intentionally planned to foster potential sales sooner than the customer expected. We all recognise and recall 20-year-old refrigerators still operating economically and effectively, while newer ones are subject to breakdowns after three to five years. Washing machines, according to repair men, are generally expected to last not much more than five years. Clocks and tape decks built into car dashboards as factory installed equipment last just until the warranty expires.

Equipment wears out because of usage, and it wears out quicker because of deferred maintenance or lack of it. Effective preventive maintenance normally extends the wear period for most equipment and products.

In both types of obsolescence, it is necessary to plan for product retirement, phase-out and disposal of items no longer necessary. Phase-out and disposition of the product normally happens once problems occur which require a decision for disposal. Product phase-out and disposal is an essential component and function of the product life cycle and must be considered early in the conceptual and advanced development phases of the system programme.

The obsolescence plan

Phase-out and disposal requires an obsolescence plan incorporating the following:

- Actions required and location of identified equipment and inventory to be phased out.
- Determination of the rate at which obsolete menus should be phased out. Although depreciation rates specified in Department of Taxation regulations should be considered, these rates of depreciation are in excess of customary obsolescent rates. Therefore, they should not be the sole determinants for management decision making.
- Verification of and measures used for disposal considerations in the product design and development process.
- Determination of what equipment and components can be salvaged or resold. Included should be determination of material and equipment to be reprocessed, converted or used for other purposes.
- Insurance that the disposal methods are consistent with the US Occupational Safety and Health Act (OSHA), the US Environmental Protection Agency (EPA) regulations, and ecological, environmental and socially responsible requirements and effects.
- Logistics support requirements necessary to perform phase-out and disposal consistent with effective waste management practices.
- Physical distribution and transportation requirements and capabilities that include handling, disassembly, decomposition, processing, required facilities and technical actions. Here the feasibility analysis, functional analysis and LSA are vital to the planning effort.
- Compliance with the design disposability criteria that occurred in the early design stage, which was specified also in the LSP.
- Assessment of logistics support effectiveness in terms of obsolescence supportability.
- Evaluation of determinants that could have brought about obsolescence sooner than planned. In this connection the following factors are suggested for review:
    - inherent reliability characteristics;
    - manufacturing defects;

- wear out characteristics;
- mean time between failures (MTBF) consistencies;
- secondary failures due to catastrophic failures (such as power);
- operated-induced failures;
- lack of maintenance-induced failures;
- mishandling errors.

It should be recognised that product failures can often be detected through built-in system self-test go-no-go indicators, but these devices sometimes require extensive, frequent calibration efforts. These devices present a visual indication of faulty components and chassis. They are particularly popular with electronic equipment. Others are based on computer programs which in themselves can be faulty. Nonetheless, they are considered essential for product performance monitoring.

An excellent technique for determining the effectiveness of obsolescence management is the maintenance task analysis that indicates a record of replacement parts, test and support handling equipment readings, description of facility requirements and special technical data instructions. Another management tool is the logic trouble-shooting diagram. The point is that effective maintenance programmes are integral to the obsolescence planning process. Also of importance is the maintenance analysis summary that presents a functional description, description of the maintenance concept and test and support equipment results. These provide quality control over the following areas of logistics support:

- Test and support handling equipment.
- Spare and repair parts usage.
- Facilities capacity and effectiveness.
- Technical data results.
- Operational and maintenance effectiveness factors.
- Maintenance concept and policy determinants.

Further discussion on the role of maintenance factors in obsolescence planning can be found in B. S. Blanchard, *Logistics Engineering and Management* (1983), Appendix B, Maintenance Analysis Data.

Waste management

Waste management in its more sophisticated form is a relatively new field. Very few engineering and almost no business administration institutions of higher education have courses in this subject. And very few business school texts refer to the subject, and no marketing or logistics courses are included in production and operations management (generally a required course for all business school students). Few references, if any, contain any discussion on waste management. The primary exception is Leenders *et al.*

(1985). In fact, the field is so unattractive that logisticians have often tried to convince college students taking various logistics courses that they might want to consider the waste management profession as a career. One young bright student, who typified the general student consensus, responded: 'What? Become a garbage man? That's what Mum and Dad said I would be if I did not go to college; besides, what would they say after spending over $16,000 per year in tuition?' Illusions, misconceptions and attitudes pervade most managers and the general public's thoughts about waste management. The result is very little interest in the problem.

Some government officials and commercial sector managers are becoming more concerned with effective solutions for the disposal of scrap, surplus, obsolete and waste materials. Disposal solutions are complex because of the size of firms and decentralised management control. Solutions for disposal are obligatory because of its impact on pollution; the need to conserve scarce, costly resources; and the renewed awareness that disposal products provide a value-added cost saving in the production of materials and services.

Savings versus annoyance

Disposal has a high degree of salvage savings potential. Some firms have come to recognise that by reducing items for disposal, reclamation and net returns they provide a significant cost benefit savings and return on investment in material.

Yet, the industrial and service sector's reluctance to capitalise on these financial returns is understandable for several reasons:

- Popular connotations that disposal and waste have no value and they constitute 'stuff' to be disposed of.
- Pervasive perceptions that products should be easily disposed of in a 'throw away' society.
- Many firms are not large enough to establish a salvage, disposal or utilisation department. Many of the larger firms which have this department or division are primarily concerned with internal reclamation within the production operation and not the sales division.
- The volatility of the price structure of scrap and waste on the spot market which complicates firms' financial planning processes.

Disposal categories

Most management theories and practices contend that effective management precludes the generation of items for disposal. In reality, every firm in the manufacturing and services sectors generates material that has to be disposed of. This material can be divided into six categories:[1]

[1] Leenders, Michiel R., Fearon, Harold E. and England, Wilbur B. (1985) *Purchasing and Materials Management*, 8th edn. (Illinois: Richard D. Irwin, Inc.).

1. *Surplus material* is that which is excess to the needs and requirements of the organisation. Handling this material may be accomplished in three ways.

   - store it until needed;
   - substitute the material for other material for work in progress;
   - sell it, usually at less than market price.

   Larger firms have the advantage of combining these options by transferring the material to another plant or office, or postponing the demand for it until sometime in the future. Most firms develop a scheme to determine whether material is surplus, such as categorising all material excess to requirements for a six-month or one-year period.
2. *Scrap* is material consisting of by-products from production, or worn machinery and tools (which normally have exceeded their depreciation schedule). Scrap is segregated by type to obtain an optimal selling price. For instance, scrap metals are separated into ferrous and non-ferrous types.
3. *Obsolete material and equipment* is usable material which could be used at a later time, but is unlikely to be so. This category exists because of improvements in new technology which usually make it more cost effective to dispose of. Obsolete material and equipment normally has a useful purpose for other firms, so it is saleable at better prices than the other categories.
4. *Waste* is material, equipment and supplies that have been altered during the production transformation process, or exist as by-products of manufacturing or services. This category is normally unfit or unusable by anyone after it is generated and has little or no resale value, with the exception of items that can be recycled or reclaimed. Although it should not exist, waste happens – usually from poor management, or production processes, carelessness or poor handling and storage. Although currently many forms of waste have become highly saleable because of the potential for reclamation for other purposes, managing disposal consumes a high degree of managerial attention which, popularly, is not accorded.
5. *Rejected products and services* are those normally rejected by quality control teams as being unfit or unsatisfactory. Some end products can be reworked to meet standards and are added to the inventory for sale. Complex high technology products, such as semiconductors, which have a high degree of rejection because of defined limits and tolerances, fall into this category. Some of the items may be sold as defective products or seconds, but the quality manufacturers are reluctant to do so because of the potential for injured product reputation for quality. Other rejected products are destroyed as in the pharmaceutical or food industries. This applies to the services sector (such as fast-food restaurants that may generate less than optimal quality in hamburgers or french fries).

6. *Safety stock* is added as a final category, not because it falls into disposal categories, but because it is often considered such by management (for immediate cost savings purposes). Safety stock consists of those components and end items set aside in inventory to be used for emergency purposes, such as responding to sudden, unexpected consumer demand, or unforeseen contingencies such as bad weather, sudden labour walkouts, or unexpected financial reversals. It is *never* to be used, except in dire circumstances, particularly to meet demand. This stock must be managed and rotated to preclude spoilage and obsolescence. It is included as one of the disposal categories because of the popular trend by many managers and CEOs to attempt to improve their cash flow by selling safety stock, and because of the increasing results of the current popular mania for mergers and acquisitions, takeovers and leveraged buyouts. New parent firms often tend to sell off safety stock again to improve cash flow and financial posture.

Disposal strategic options

The methods for disposal become more limited daily because of increased legislation due to waste management negligence and the worldwide public outcry for improvements in ecology and environmental protection. The firm also has to maintain a profit and bear its shareholders (if applicable) in mind.

The alternatives to the disposal of material and supplies require intense top level management. There are relatively few options. Among them are:

- *Use 'as is' within the firm.* In this case, material and supplies are transferred from one department or plant to another. They may be substituted for other requirements, or modified to use for other purposes.
- *Reclamation for intra-plant use.* Material and supplies used for other purposes or modified to fill the need of another department or division. Some scrap may be recycled.
- *Sale to another firm.* Material and supplies may be sold 'as is' through direct sales or through a broker. Some economic return on the material is then realised. Surplus, obsolete and even safety stock most often fall into this option.
- *Return to the vendor.* Although most vendors will not accept return of material and supplies, the steel mill industry stands out as a notable exception for reclamation of steel scrap.
- *Sale to a dealer or through a broker.* A scan of the yellow pages of the telephone directory will reveal a dramatic increase in the number of dealers and brokers who deal in scrap. With renewed public interest in ecology and pollution prevention, it is becoming more attractive and lucrative to buy and sell scrap. The scrap market is becoming more competitive and less volatile in prices.

- *Dumping.* Many larger firms have adequate real estate to provide for on-site dumping of material, although local and state departments of environmental management are beginning to look askance at this practice. Other firms have no place to dump their waste, and several resort to middle-of-the-night operations in remote land areas and rivers. Others have been polluting rivers and streams with hazardous waste and sewerage since the origins of the firms. Landfills in most areas are now over capacity

The current waste disposal situation

As previously indicated, each person in America generates an average of 1,000 pounds of garbage every year. Landfills are over capacity, and alternate solutions are overdue. Ineffective waste storage, transportation and disposal have generated gigantic problems in America and Europe.

According to Kenneth Brooks and Anne Watsman, one alternative solution is to develop methods to burn garbage to produce steam to generate electricity.[2] Called resource recovery plants (RRPs), the concept has expanded rapidly, and over 100 plants have been built. The primary contractor is Combustion Engineering Inc. (C-E), which provides mass burn and refuse-driven fuel (RDF). Innovative as this technology may be, landfill owners complain that the operational costs for these systems exceed their payoff factors.

Although recognised as a long-standing problem, asbestos fibres continue to plague the workplace. Today, managers still appear to remain silent and refrain from telling their employees of the asbestos hazards to the firm. Every nation in the Western world has different policies and practices for handling the asbestos problem.

Landfills, both municipally and privately operated, in addition to overcapacity problems, also create pollution risks by producing methane gas from organic material that is buried. Many have experimented without significant success with one or more of the options to resolve the problem such as continued burial, recycling, composting and incineration.

Hazardous chemical waste disposal is an increasing problem because of transportation and inadequate disposal services, including where to dump the waste. Compounding this issue is recent US Congressional legislation that provides for generators of such waste to be liable by the Government and the courts as responsible for total cleanup costs under the doctrine of joint and several liability. This means that chemical-producing firms can be held liable indefinitely, and implies that firms should institute mechanisms and procedures to track contractors who remove and dispose of this waste to ensure that the vendor actually disposed of it properly.

---

[2] Brooks, K. and Watsman, A. (1986) 'Garbage Disposal: to burn or not to burn', *Chemical Week*, 11 June, **138** (34), pp. 44–47.

Ground water pollution, non-point source pollutants and hazardous waste disposal are especially difficult, according to a journal article in the May 1988 issue of *Chemical Week*. Lysle D. Helsing points out that for the past few years under the Reagan Administration there has been a significant decrease in federal funding to assist with this problem. States and local communities have had to pick up the responsibility at a time when they, too, suffer significant deficits due largely to the impact of the Tax Reform Act of 1986. Helsing estimates that it would take $76.2 billion in capital investment to provide adequate wastewater treatment by the beginning of the twenty-first century. Federal expenditures for water pollution control dropped by 31% from 1978 to 1984.

The ILS solution to waste disposal

For all of these situations, the use of the ILS logistician has been notably ignored. Most of the problems have not been attacked with the services of a logistics professional who could have provided optimal initial planning, funding requirements and recommended controls to ensure that waste disposal was effectively managed. As a result, no one has integrated all the various elements of logistics support. The management of waste disposal needs ILS – a composite of the requirements to support the waste disposal system through its life cycle. For waste disposal management the ILS logistician can provide professional assistance and leadership through maintenance planning, test and support requirements, facilities planning, personnel and training, handling and transportation planning, logistic support resource funding requirements and management information. The ILS manager should have a position high enough in the waste disposal organisation to plan, monitor and provide management control. He can assist the organisation in meeting logistic support objectives, milestones and procedures to meet waste disposal task objectives. The ILS manager is a professional who can provide project-oriented direction for waste disposal, while other lower-level managers provide technical direction according to their specialities.

Conclusion

In summary, waste disposal is a problem of severe magnitude, long overdue for solution. Managing the toxic and polluting effects of waste must become the highest of priorities for the 1990s. Immediate action is required by private industry, government and private citizens. Creative financing to initiate and upgrade research and development has to be implemented. Government agencies and private firms would be wise to utilise the services of the ILS logistician to implement the planning, managing and control functions necessary to bring optimal waste management practices to fruition. Bureaucratic red tape delays and apathy are no longer acceptable, because the urgency of the situation demands immediate action. One need only recall the drought summer of 1988 in the United States, that demonstrated the current damage to the earth's ozone layer, to come to the conclusion that everyone has a mandate for change. The logistician, who systematically effects change, is key to this process.

Chapter 13 ILS SYSTEM SUMMARY

The previous sections of this book covered the integral elements necessary to ensure that logistics is accomplished in a timely and effective manner. The elements that have a direct impact on the success of logistics and the elements that drive logistics itself were described. Nine specific logistics elements and how they interract with one another were discussed. But what was not done, and was left until last, was to integrate the functions associated with logistics and those that drive logistics. The purpose of this last chapter is to give the reader an understanding of how all of these elements are tied together. The hope is to show from a systems level why logistics is important and how it should operate to ensure that the goals that were discussed earlier are achieved. Specifically, logistics should provide the maximum availability of parts and personnel for consumer acceptance at the minimum life cycle cost possible.

During the initiation of the programme, the project or acquisition manager should identify what his goal is and establish the minimum requirements for effective ILS management. It is extremely important during this phase that an ILS manager be designated as soon as possible so that he can assume the necessary ILS management responsibilities and ensure that all of the special aspects of the programme can be quickly identified and organised so that ILS flows well and can be easily integrated into the programme. If there are any special logistics support requirements, they should be identified early so that they can be handled as part of the programme rather than becoming an adjunct to it.

The ILS manager, therefore, must establish the ILS planning documentation that will be used, frequently reviewed and updated to ensure that the programme is completed on time and that it solves the problems necessary to supply the appropriate support requirements. This is generated through the work breakdown structure and the utilisation of task-oriented requirements statements, which become part of the planning documentation established by the logistics manager. The initial milestones are developed for the overall programme so that everything works as a system rather than as individual items. The life cycle cost parameters and other aspects associated with costing in the programme are also identified so that the mechanisms and tools are in place to assist in determining the impact of trade-offs and the types of studies that need to be performed and approaches that are to be taken to ensure minimum life cycle cost.

During this early phase, the logistics manager works in conjunction with the systems engineer to prepare the appropriate documentation (specifically, the maintenance approach by which the product is going to be designed). He could use the contract definitions that were supplied by the procuring activity or the inputs that were received from the field service personnel, who have to know what the customer wants and how best to minimise the impact on him, whether the product is intended for manufacturing or personal use. Once this information has been defined and the work has been established, profiles become a secondary aspect, but are used to control the programme to ensure that there will be no overruns. When this has been completed, and the tasks have been identified, the work necessary to initiate a programme commences.

The logistics manager's requirements, therefore, are reflected in what was included with the proposal or information that he obtained during the early initiation stages of the programme. For example, a contractor who is proposing a set of new designs for a prime contract would be required to prepare a response to a solicitation. It is extremely important that he form a proposal team consisting of the key personnel necessary to ensure that the material to be supplied as part of the proposal is properly organised and implemented. For ILS the manager must coordinate all of the information that is available to him, either using historical data obtained through the LSA data base or using the field service personnel's historical records to determine how best to handle the specific product.

All of this information becomes the basis for future design. It is important that the logistics engineering manager recognise that the material he supplies is a basis for the work he will do once the award has been made. The development of an ILS plan, commonly called an integrated support plan, by the contractor is important because it is an extension of the work breakdown structure to show how the logistics manager includes his tasks into the total programme. It also establishes the tools that will be used to analyse the alternative approaches when designing the new product. Detailed recommendations will be included in the plan to reflect specific content, milestones and cost projections necessary to ensure that logistics becomes an integrated function.

Once the programme has been awarded and the initial design aspects have been satisfied, the logistician proceeds directly into full-scale development of the product. Let us call this the performance phase, so as to identify exactly what the ILS manager's requirements are. Prior to this, or during the concept of design and validation phase, the only aspects the logistics manager needs to be concerned with are the institutionalisation or the finalisation of a maintenance approach, the inputs to the reliability studies and the establishment of support criteria. Thus the reliability engineer can quickly assess his impact on the design and what he must do to assist the logistics manager in ensuring that reliable components are used as required and included in the cost analysis to help reduce the total cost to the consumer.

During the performance phase, the ILS manager is required to supplement the various ILS plans as the development proceeds. He will co-ordinate the ILS system with the hardware development so that the two work in parallel to ensure that the logistics engineer measures the degrees of change appropriately, and reports them to the design engineer so that alternative designs and approaches can be quickly modified and implemented. This again requires team meetings and work groups to maximise the input and minimise the impact on the programme. By this approach, communications are maintained with less effort because everyone is part of the recommended design alternatives. During this phase, the configuration manager becomes a major player because the specifications and the other guiding documents become part of a design programme freeze. Once the documents have been approved, any changes relative to the specific document need to be appropriately identified, documented and recorded so that as the design proceeds into the production phase, we have established an audit trail.

Last but not least is the identification of critical support problems and the actions that need to be taken to ensure that these are resolved. The importance here is tying it into the team meetings and work groups to minimise the time required to resolve these problems. Special *ad hoc* committees should be set up to ensure that the problems, when identified, are evaluated to determine the impact they have on the specific programme and, if the problems cannot be resolved, that higher authority is brought to bear so that a quick determination can be made.

Coming into play at this point are three additional people within the logistics support team. They will be responsible for assisting the design engineer and establishing the types and locations of the lowest repairable units. Along with the maintainability engineer, the maintenance engineer is of prime importance in assisting both the design and maintainability engineers in determining how best to locate and isolate specific problems that may arise within the design. This will complement the work that will eventually be needed from the training and support and test equipment personnel because they will be responsible for identifying and determining the support and know the difficulties involved in training people once the equipment is distributed.

The maintenance engineer becomes a prime factor in the design phase, and is normally on-line and working with most of the other team members. This does not mean that he is involved in the design phase on a full-time basis; rather, that during the period in which he is involved in the design, he devotes all of his time to it. He also participates in the test and evaluation later on in the design phase to ensure that his maintenance approaches and specific diagnostic routines have been satisfactorily performed and implemented. This is done so that problems can be rapidly identified, localised and isolated as they occur. This approach will give the maintenance technician the ability to solve the problem and demonstrate achieving the MTBR that was established for the contract.

This work by the maintenance engineer, in conjunction with the maintainability and design engineers, is performed between the PDR and the CDR. This is when the drawings become available and are structured so that production information can be finalised. The maintenance engineer also works in conjunction with the reliability engineer who is attempting to finalise his reliability estimate so that the maintainability and maintenance engineer can incorporate the component parts, assemblies and so on into the appropriate locations to help modify the impact of a failure. It is also important for the reliability engineer to quickly identify the need to establish redundancy in the design. If done too late, it will have a dramatic effect on the designer's ability to satisfy the requirements of the contract.

The next player in the logistics management responsibility is the logistics support analysis engineer. He is responsible for ensuring that the various studies performed during this phase of the contract are prepared and that data become a relevant part of the total design. The LSA record base, Fig. 13.1, therefore has been included as part of this discussion; its use is similar to that of a circular slide rule. It can be used by the logistics manager to help establish which tasks need to be performed during which phase of the contract. The flow of the circular slide rule works in a clockwise manner, where everything begins with task 101 and concludes with task 501, but it becomes a sequence of events processing through the preconcept, concept and D&V through full-scale engineering development and production. (The tailoring is discussed in Chapter 11). Preliminary efforts of the training element manager, the test equipment manager and the technical manual manager begin during this phase. Information entered into the logistics support analysis data record base becomes the material that is used by these element managers to prepare the tasks that they will be required to accomplish. These element managers support the full-scale test and evaluations performed at the conclusion of this phase.

The ILS effort is an iterative effort, and must be considered and handled on that basis. The trade-offs that were performed in earlier phases should be looked at again to determine whether they are still applicable. If they are

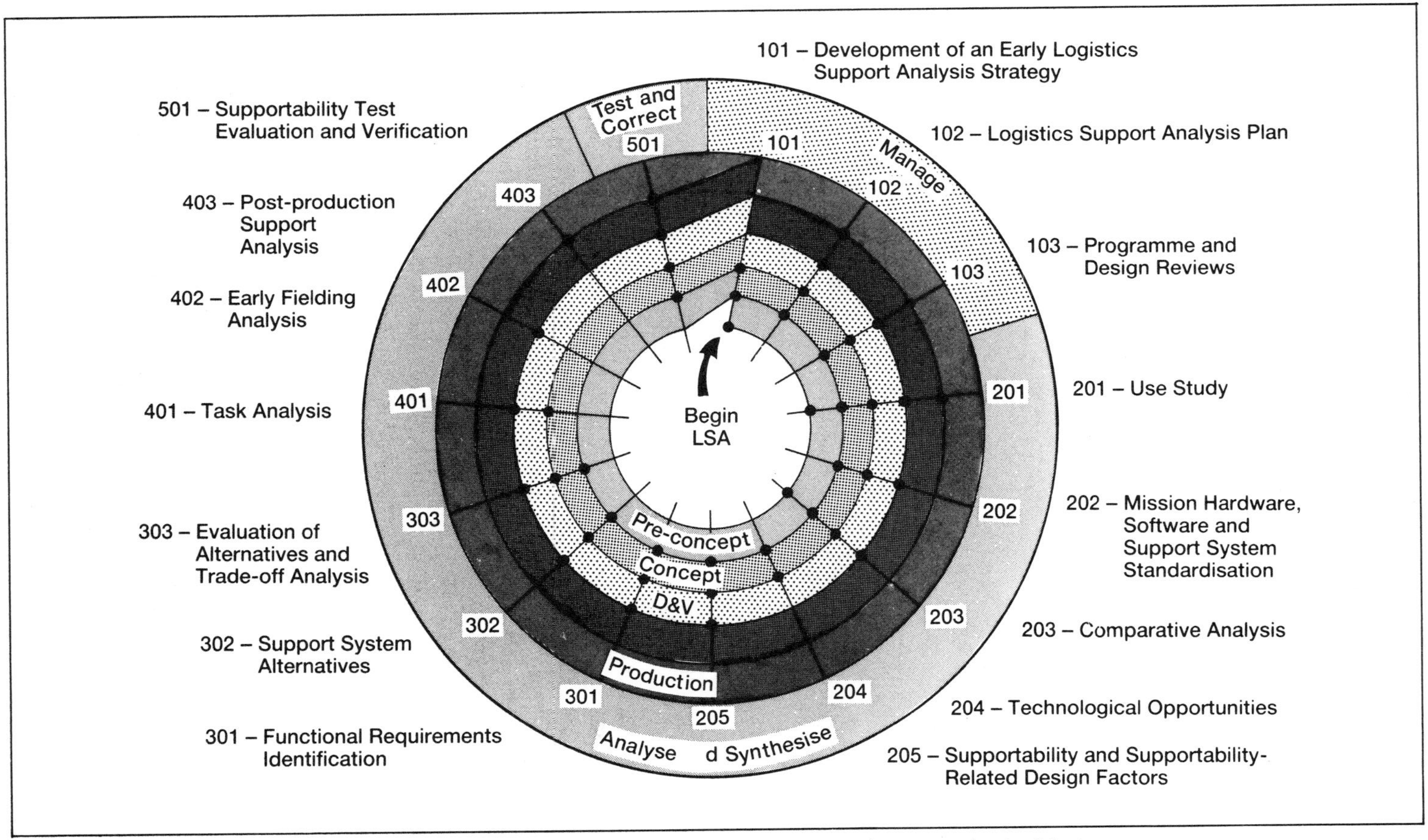

Fig. 13.1 Relationship of logistics support analysis to ILS and system engineering. D&V, demonstration/validation.

not, they should be reviewed and re-run to determine whether a more applicable approach is necessary in this phase of the contract. The trade-offs will have a direct impact on the tasks that will be performed by the LSA manager and other resource managers within the logistics environment. Refinement becomes a key point during this phase because, as this phase is concluded, most if not all of the logistics resources will be defined and implemented. The specific efforts will be concluded during the production phase of the acquisition life cycle. All of the problems that were identified should be resolved by the conclusion of the performance phase. Those that still remain must be resolved prior to testing and evaluating the equipment. If this does not occur, the probability of a successful test decreases. The logistics engineer should work closely with his element managers to ensure that outstanding problems are quickly resolved.

Once the product has been prototyped and is ready for testing, the entire support team becomes part of the community that is used to evaluate the manufactured product. They should ensure that the appropriate technical documentation is in place; specifically, the test procedures and plans, and the necessary failure-reporting actions required to report the problems that occurred during each phase of the testing cycle. This failure-reporting tool becomes the mechanism by which the logistician and the design engineer report back to the systems engineer the problems that are being experienced and the attempts to determine trends during the test cycle. If these trends appear, it is absolutely imperative that the team evaluate their impacts to determine whether a latent defect has been built into the product. If that is the case, corrective action must be implemented immediately so that the test can be completed to the satisfaction of either the customer or the manufacturer to ensure that once the product is distributed, it satisfies the ultimate goals.

All the tasks being performed during this time have been placed under configuration control. Any changes associated with drawings, manuals and any other technical documentation that was discussed in Chapter 7 must be coordinated through the configuration management office to ensure that an audit trail is established for all the changes that will need to take place.

Once all the tests have been performed and the product has successfully met the requirements as stipulated by the functional specification in the support requirements documentation, it then becomes a production item. This leads to the implementation phase of the acquisition cycle. In the management section, it was stated that the logistician should assume responsibility for the production cycle because 95% of the efforts have a direct impact on what he is required to do.

During the production implementation phase of the final piece of equipment, the ILS manager becomes more dynamic. His responsibility is to ensure that all of the pieces to the puzzle are properly put together and that they satisfy the ultimate goal. He is also responsible for the follow-up work on the various tasks associated with ILS. Specifically, he needs to ensure that the appropriate designs and all the documentation associated with them will support the generation of any specific test equipment needed to support the product that is going to be used. Factors that need to be considered as part of the overall design are no different from those that were defined for the prime item development of the product that these pieces of test equipment are going to support. They must be looked at as automated versus manual. The reliability of the tools, or the monitoring or the checkout equipment, must be similar and should be equal to or better than the reliability of the equipment to be tested so that the test equipment does not become part of the problem.

The test equipment must be easily calibrated so that the service personnel (or whoever is involved) can ensure that it meets the local standards as well as being traceable back to a National Bureau of Standards. Test equipment becomes part of total logistics planning during the full-scale production phase. It is important that test equipment be considered early so that it becomes part of the total procurement. The test equipment should also be fielded in conjunction with the final end product so that when personnel in the field receive the end product, they are able to take care of the bits and pieces as required. It should fit all levels of support, from the organisational level back through to the manufacturer's facility.

It is also important that supply support activities and tasks be planned early to be part of the implementation phase. Supply support (specifically, the spare or repair parts, the supplies and inventories or repairables) should be predetermined so that the state of readiness of the equipment and the supply lines can be filled up to support any kind of distribution requirements that are needed by the user. Spare parts should be procured along with the prime item development equipment so that the contractor can take advantage of larger purchases including spare and repair parts. The logistician should also plan the various aspects that are a result of the LSA. He considers whether to repair or discard the failed part. Emphasis should be placed on the facilities receiving the part and be capable of maintaining the supply readiness demand.

The spare part should be available at the time of use. In many cases, they are the last to be considered and the first to bring the equipment down because they are not there when needed. The logistician is responsible for ensuring that the part is where it is supposed to be when it is needed, and is operational, so not affecting the capabilities of the equipment. The supply support works in conjunction with the other aspects of manufacturing and is tied into the total manufacturing base. It is a continual effort and proceeds through the entire implementation phase and continues beyond if required to support the equipment after the manufacturer has finished his production run.

Also at this time there is the need to ensure that the physical distribution, transportation and handling functions have been studied and the necessary actions to ensure the capability to transport, preserve, package and handle all equipment and supply items have been taken into consideration. This becomes an extremely important facet of the logistics engineer's responsibility. It means making sure that the appropriate documentation has been prepared so that it defines the transportation criteria, the handling equipment and procedures, and the packaging and preservations needed to ensure that the equipment, whether it be test equipment, prime item development equipment, or spare or repair parts gets to its destination and is operational when it is unpacked and installed. Items that need to be looked at include cost and the location and duration of transportation, all of which have a direct impact on how best to get the product to the user.

Transportation handling and the other supporting activities associated should not be overlooked. This task (PD&T) is described in Chapter 7. It should not be overlooked and should be properly coordinated as early as possible.

Technical data that will be generated during this phase consist of the data that will be needed to support the efforts associated with the equipment once it has been distributed. The data to support this information are drawn from the logistics analysis record database that was generated during the earlier phases. The LSAR database helps in the generation of the final documentation of the operating and maintenance instructions. It establishes the provisioning and facility information requirements base from which all the other facets of the programme need to be derived. It includes the data involved in the design and the actual support functions associated with the configuration management databases. It is also the mechanism used to document failures which occur during the implementation and production phase. The process maintains its own historical record for the determination of trends and other analyses that may need to be performed by the logistician once the equipment is in use.

As part of the full-scale engineering development phase, the need for personnel training was discussed. During this phase, training courses were prepared and preliminary training was presented to personnel who were going to perform the user's side of all of the tests. As a result of the test base and the initial training, the personnel and training course material would be updated and prepared to meet the needs of the customer or procuring activity. The training courses will also be incorporated into the evaluation of the test equipment.

Repair procedures are needed to maintain the test equipment so that it can be used to fix the prime equipment as discussed earlier. Therefore, not only is training required for the prime equipment but it is also essential for the new test equipment. The material for the training programme is finalised and submitted to the procuring activity or the customer. They may opt to allow the manufacturer to continue his training programme techniques until the user or customer establishes his own capabilities. From the Government's standpoint, it is normal for a series of courses to be continued until the cadre has been satisfactorily instructed on the applications and the use of material in hand. Then the contractor or field service personnel review the training programmes, correct the areas that are needed, and leave the facilities because the rest of the training will be performed by the end item user. All of this material is sequentially prepared in accordance with the overall schedule of the programme. The training courses also help demonstrate that the maintenance and operating instructions were correctly prepared and any errors found during the earlier phases or the acquisition cycle have been corrected. The validation phase is no different from that which was performed during full-scale engineering development with the documentation finalised and submitted as part of a total package with the prime item development product.

The performance or deployment phase constitutes a portion of the entire acquisition or life cycle of the product. During this phase of the equipment cycle, the ILS manager is responsible for serious and difficult tasks necessary to support the product that he has just deployed. The first task is to fine tune all of the start-up requirements that are necessary to ensure that the product is integrated into the consumer's life cycle or into the infrastructure established by the procuring activity. He must ensure that, from the standpoint of ownership, all of the materials associated with the product have in fact been implemented and are in place.

This becomes a very critical factor because if the user believes that he is not receiving the material or the support documentation necessary to ensure proper operation, he will be dissatisfied and could create problems for the manufacturer on future design applications. The logistics manager is responsible for establishing and maintaining a systems approach for defining problems versus identifying symptoms instead of the problems themselves. A very clear cut approach needs to be used so that the appropriate feedback is established and the parameters and methods for this are in place at the time of deployment. The field service personnel are responsible for ensuring this. They will interact with the end user, and will be responsible for trying to avoid any catastrophic problems with the persons using the final product.

The logistician must also avoid 'fire drills' which are sometimes both inappropriate and very non-productive. They are a way of reacting to problems associated with the product without having fully detailed information on what the problems really are and what their cause is, and without a good appreciation of the customer's point of view. 'Fire drills' turn out to be a real problem for management because it costs money to react. Logisticians do not want to be reactive, but extremely responsive. The feedback obtained from the field service personnel should be the basis from which to attack all problems perceived by any of the people involved in the use of the product just distributed.

The ILS Management Flow Chart* shows the flow that the ILS manager must maintain and which must be consistent with the design approach. This flow diagram has been constructed to give the reader an understanding of how all the pieces of the puzzle fall into place. The diagram covers all the logistics support elements and their interfaces with some of the major

* Provided separately with this book.

units. At the top is the generation of the programme management offices and what is necessary to ensure that the proper design and controlling packages have been integrated. Second is the system design management flow, reflecting the ideas and requirements to ensure that the design satisfies the specific programmes.

The Product Life Cycle Technical Activities Chart* augments that portion of the systems or design engineering activity, identifying the systems life cycle technical activities covering all of the tasks associated with a specific programme. It also covers the specific design reviews and how ILS is integrated into the total package. The major flow diagram, from a logistics management aspect, takes the last box (ILS) of the system life cycle technical activities, and expands it into what has been included in the document. It is reasonably self-explanatory, and all the actions required to support this have been defined. (Note that this was originally prepared for use by the Government agencies for a generation of military equipment, but is equally applicable to the commercial situation.

Summary

The above attempts to ensure that the logistics manager has a good appreciation of everything that must be accomplished from the onset of a programme through its removal once it has completed its life. Proper balance must be structured between the operations and requirements of the corporation so that support and performance are properly organised, are equal to technical performance, and are maintained throughout the life cycle of the equipment.

* Provided separately with this book.

APPENDIX A. ILS RESPONSIBILITIES

Assistance the programme management office must supply to ILS

- The rights to establish a charter under the programme guidelines
- Programme risk analysis
- Total involvement in the review of all programme requirements/ objective/activities (depending on the scope of the programme)
- Establishment of proper communication
- Participation in all of the discussions leading to overall programme decisions
- Overall system schedule/projections
- Cost projections/constraints
- Product planning
- Work breakdown structure
- Assignment of responsibility and authority

Assistance ILS must supply to the programme management office

- Schedules/cost of programme(s)
- Charter under which ILS will work (planning documentation controls/ directives)
- LCC evaluation of total system cost (proposed)

- Support in overall programme decision-making
- Support in all programme reviews (internal/contractor)

Assistance systems engineering must supply to ILS

- Product plans/concepts/specifications
- Hardware interfaces (proposed) (inter and intra)
- Mission requirements/constraints
- Systems effectiveness criteria
- Omission profile
- Systems modelling
- Contigency and sensitivity analysis
- Risk analysis
- Product volume, weight, size
- Environmental conditions
- Scenarios
- Figures-of-merit
- Testing concepts/requirements
- Systems approaches
- Duty cycle(s)
- Product operational support to technical publications
- Critical engineering paths
- Engineering design trade-offs

Assistance ILS must supply to systems engineering

- Performance parameters (reliability, availability, maintainability/probability of mission success)
- Definition of maintenance support concepts for all echelons
- Definition of levels of skill requirements (months)
- Levels of self-test capabilities
- Training requirements
- Facilities criteria
- Supply support criteria
- Transportation and handling criteria
- Technical information criteria
- Support equipment criteria
- Field support criteria
- Life cycle cost predictions
- Spares requirements

- Trade-off models (system versus support requirements)
- Installation requirements

Assistance the design engineer must supply to ILS

- Stress evaluation – stress/temp/duty cycle (application)
- Quantities
- Hardware design approach – implementation of systems specification
- Incorporation of maintenance support requirements
- Power requirements/facility requirements
- Weight, shape and size of equipment configured
- Engineering change request
- Functional block diagrams
- Subsystem/equipment specifications
- 'Unit' test requirements
- Schematics
- Critical paths (reliability analysis)
- Suggestion for spares requirements
- Trade-off approaches (models)
- Design review

Assistance ILS must supply to ILS control

- Packaging and packing transportation plans
- Spares programme plan
- Reliability/maintainability plans
- Math models for evaluation of reliability and maintainability
- Task and skill analysis
- Maintenance concept/plan
- Repair plan
- Support goals/criteria
- Subcontractor requirements
- Preliminary personnel requirements
- Preliminary training requirements
- Maintenance engineering analysis/logistics support analysis
- Facilities requirements
- Field support requirements
- Technical manuals/documentation requirements
- Logistics support plan
- Support equipment requirements

- Configuration management/change control
- Equipment layout/printed circuit board/assembly
- Trade-off studies
- Long-lead items
- Critical support path
- Design redundancy
- Support to training
- Support to technical publications
- Support equipment requirements
- Field support training (prototype equipment)

Assistance ILS must supply to the design engineer

- Parts criteria (preferred parts lists) – acceptable stress levels/temp levels/usage levels
- Predictions allocations/analysis reliability and maintainability
- Test requirements
- Skill level requirements – enforcement of system level requirements
- Maintenance concept requirements (repair/discard) levels of maintenance
- Packaging criteria
- Drawing control (configuration management)
- Test support requirements (special equipment/general purpose)
- Cost contraints (LCC evaluation)
- Logistics support analysis – to support design concept results of FMEA
- Data bank/control/distribution/data collection analysis
- Design review support
- Funding requirements
- Calibrations requirements
- LCC analysis
- Preferred parts control

APPENDIX B. CLASS I ENGINEERING CHANGE

(Adapted from MIL-STD 480, entitled *Configuration Control: Engineering Changes, Deviations and Waivers*)

An engineering change shall be classified as Class I when one or more of the factors listed below is affected:

- The functional or allocated configuration identification (FCI or ACI).
- The product configuration identification (PCI) as contractually specified (or as applied to government activities), excluding referenced drawings, specifications, listings of computer program instructions and actual data values.
- Technical requirements below contained in the PCI as contractually specified, including referenced drawings and specifications:
  - Performance outside stated tolerance.
  - Reliability, maintainability or survivability outside stated tolerance.
  - Weight, balance, moment of inertia.
  - Interface characteristics.
- Non-technical contractual provisions:
  - Fee.
  - Incentives.
  - Cost to the procuring activity.
  - Schedules.
  - Guarantees or deliveries.

- Other factors:
  - Client-furnished equipment.
  - Safety.
  - Electromagnetic characteristics.
  - Operational, test, or maintenance computer programs.
  - Compatibility with support equipment, trainers or training devices/equipment.
  - Configuration to the extent that retrofit action would be taken.
  - Delivered operation and maintenance manuals for which adequate change/revision funding is not on existing contracts.
  - Pre-set adjustments or schedules affecting operating limits or performance to such extent as to require assignment of a new identification number.
  - Interchangeability, substitutability or replaceability, as applied to CIs, and to all subassemblies and parts of repairable CIs excluding the pieces and parts of non-repairable subassemblies.
  - Sources of CIs or repairable items at any level defined by source control drawings.
  - Skills, manning, training, biomedical factors or human engineering design.

NOTE: In the above definition of a Class I engineering change, the words 'excluding referenced drawings, specifications, listing of computer program instructions, and actual data values' in the second item shall not be interpreted to exclude these items prescribed directly in a contract to define contract line items. Other drawings, specifications, computer program instructions, and actual data values, whether referenced in documents or listed on associated lists, are excluded from all of the above, except the first item.

APPENDIX C. MAINTENANCE PLAN

(Adapted from Uniform Data Item UDI-L-1241S, Naval Supply Systems Command, 72 May 15)

The maintenance plan is the keystone of subsequent efforts to provide integrated support by providing a coordinated centralised technical database for use in all subsequent logistic resources acquisition planning and implementation. This document portrays decisions pertinent to detailed maintenance requirements, relationship to specific corrective and preventive maintenance tasks to the applicable level of maintenance and designation of resource requirements together with their estimated rates of consumption.

Maintenance plan format

The following format shall be used as a guide for developing the maintenance plan. The format and content may be modified or expanded to adapt to the specific needs of the programme.

The maintenance plan shall be prepared, and bound in looseleaf form to facilitate changes and revisions. The plan shall be arranged in the following sections:

I Introduction
II Maintenance concept
III Support and test equipment requirements summary
IV Supply support requirements summary
V Personnel and training requirements summary
VI Technical data requirements summary
VII Facilities requirements summary
VIII Transportation and handling requirements summary

Appendix A Interim master repair list
Appendix B Organisational level requirements
Appendix C Intermediate level requirements (if appropriate)
Appendix D Depot (user/manufacturer) requirements
Appendix E Maintenance task requirements summary
Appendix F Security requirements summary

Maintenance plan content (by section)

Section I

Introduction

1. *Scope.* Describe the contents of the plan.
2. *Planning data.* Provide a summary of basic planning data used in the development of the maintenance plan.
3. *Analysis technique.* Provide a narrative description of the approach, methodology and analytical or other techniques used in determining the quantitative data developed and presented in the plan including, but not limited to, usage data, repair cycle times, cost data, rotatable pool size, support equipment utilisation factors and maintenance man-hour requirements.

Section II

Maintenance concept

Maintenance concept. A narrative description of the manner in which the system will be maintained, by major equipment function and by maintenance level. This will provide an introduction to the type of maintenance actions required and a brief description of the overall concepts by which the system and equipment will be maintained.

Section III

Support and test equipment (S&TE) requirements

1. This section provides the technical and procurement data necessary to plan for and acquire and test equipment.
2. All Special Support Equipment (SSE) and General Support Equipment (GSE) required to support the system or equipment should be listed or referenced in this section. When GSE requirements are not available in the early phases of the system or equipment life cycle, then the proposed requirements should be described and updated with the firm requirements as they become available. Describe actions taken to assure maximum use of all GSE already in inventory. Elaborate on any proposed built-in test features and the acquisition, use and update of the test tape programmes for isolation or check-out. Provide a list or reference of all GSE required to support the system or equipment, describe functionally, and include, if appropriate and available, the following information for each end article:

(a) Name and nomenclature, part number, size, weight, cost, military designation, etc.
(b) Photograph or sketch.
(c) Common or peculiar designation.
(d) Maintenance level at which used.
(e) Reference to Support Equipment List (SEL), Allowance Parts List (APL), etc.
(f) Applicable calibration requirements.

Section IV

Supply support

This section will identify the spares (including high value and high failure repair parts), the maintenance levels at which each replacement or repair will be accomplished and an expression of frequency with which such replacement or repair is expected to occur. Included (as Appendix A) shall be an Interim Master Repair List (IMRL) based on the contents of the Recommended Top-Down Breakdown List prepared as part of the LSA programme. This list will provide basic key information as indicated below.

(a) Part number.
(b) FSN, if known (for military equipment only).
(c) Reference designation, if any.
(d) Nomenclature.
(e) Total quantity per system.
(f) Lowest designated level of repair.
    (1) O – Organisational.
    (2) I – Intermediate.

(3) D – Depot.
(4) X – Discard.

(g) Unit price.
(h) Estimated repair cycle times for repairable items coded I or D.

Section V

Personnel and training section

This section defines the personnel and training requirements needed to operate and support the item on which the analysis is being conducted. The data to be included will be sufficient to permit complete identification and quantification of all personnel requirements; development of personnel and training programmes; training requirements; and resources for both operation and maintenance. These requirements include, but are not restricted to:

(a) Manpower requirements.
    (1) Commercial/military personnel.
    (2) Industrial personnel.
(b) Training requirements.
    (1) Training concept.
    (2) Initial training operations.
    (3) Subsequent training operations.
    (4) Technical training equipment/devices.
    (5) Training software.
    (6) Construction and/or building rehabilitation.

Section VI

Technical data

Technical data plan (logistics). This plan will contain an identification and description of all information (publications, engineering drawings, maintenance aids, technical maintenance standards, etc.) necessary to enable personnel to operate the equipment and accomplish planned or corrective maintenance.

Section VII

Facilities

1. All support operational, maintenance and training facilities shall be described in terms of real estate, site improvements, physical structures and supporting utilities.
2. Typical facility listing may consist of complete space and/or support equipment requirements (e.g. a 'clean' room for equipment test, repairs and assembly; a degreasing facility for breakdown and reassembly).

Section VIII

Transportation and handling

This section identifies the packaging, handling, storage and transportation requirements and their support requirements.

Section IX

Installation and checkout support requirements

This section identifies and delineates the Installation and Checkout requirements.

Description of appendices

1. *Appendix A. Interim master repair list (IMRL).* To be supplied by contractor.
2. *Appendix B – Organisational requirements.* This appendix shall contain four tables listing the assemblies of the system which are repairable at the organisational level and the documentation, support equipment and personnel required to accomplish these repairs. Each of the four tables is discussed below:

(a) *Table B-1, Organisational Repairable Item List.* Table B-1 contains the list of assemblies which are repairable at the organisational level.
  - (1) *Reference Designation.* Circuit designation of the assembly in the equipment.
  - (2) *Nomenclature.* Name of the assembly.
  - (3) *Part No.* Part number of the assembly.
  - (4) *Total Qty/System.* Quantity of this assembly type in the equipment.
  - (5) *Maintenance Requirement Number(s).* Maintenance requirement Number(s) from applicable LSA worksheet which necessitates organisational level repair.

(b) *Table B-2, Organisational Documentation.* Table B-2 lists all of the documentation required by the organisational unit to accomplish repair.
  - (1) *Reference Designation.* Same as Table B-1.
  - (2) *Schematic.* Drawing number(s) of any schematics required.
  - (3) *Wire List.* Drawing number(s) of any wire lists required.
  - (4) *Test Instructions.* Identification number of any test instructions.
  - (5) *Additional.* Any other documentation which would assist the organisational unit accomplishing repairs such as exploded views, mechanical drawings, etc.

(c) *Table B-3, Organisational Support and Test Equipment.* Table B-3 shall list all of the support and test equipment required to accomplish authorised maintenance at the organisational level. The following data shall be provided for each item listed:

(1) Name.
(2) Part No.
(3) Quantity Required.
(4) Price.
(5) FSN.

(d) *Table B-4, Organisational Personnel.* Table B-4 contains the skill level required by organisational personnel to accomplish repair. In addition, the list shall include an approximation of the number of assemblies per year which require organisational repair and any additional training which will be required by organisational level personnel.

3. *Appendix C – Intermediate level requirements.* This appendix shall contain six tables listing the assemblies of the system which are repairable at an intermediate level; the documentation, support equipment, personnel and facilities required to accomplish these repairs/overhaul; and for supply support. Each of the six tables is discussed below.

(a) *Table C-1, Intermediate Repairable Item List.* Table C-1 contains the list of assemblies which are repairable on the tender of base and will include the following additional description for each repairable item:
   (1) *Reference designation.* Circuit designation of the assembly in the equipment.
   (2) *Nomenclature.* Name of the assembly.
   (3) *Part number.* Part number of the assembly.
   (4) *Total Qty/system.* Quantity of this assembly type in the equipment.
   (5) *Maintenance requirement number(s).* Maintenance requirement number(s) from applicable LSA worksheet which necessitated tender or base repair.

(b) *Table C-2, Documentation.* Table C-2 shall list all of the documentation required by the tender or base to accomplish disassembly, test, repair/overhaul, assembly, inspection and calibration of the modules and test equipment. The list information shall be keyed to the equipment by reference designation.

(c) *Table C-3, Intermediate Support and Test Equipment.* Table C-3 shall list all of the support and test equipment required by the tender or base to accomplish intermediate level maintenance. The following data shall be provided for each item listed:
   (1) Name.
   (2) Part No.
   (3) Quantity required.
   (4) Price.
   (5) FSN (as applicable).

(d) *Table C-4, Personnel.* Table C-4 contains the skill levels required by the tender or base personnel to accomplish disassembly, repair/overhaul, assembly and test calibration of each repairable item. In addition, the list shall include an approximation of the number of assemblies per year which require tender or base repair and any additional training which will be required by tender or base personnel.

(1) *Training requirements.* Any unusual training requirements needed by tender or base personnel to accomplish the repair.

(e) *Table C-5, Repair/Overhaul Facilities.* Table C-5 will specify the tender/base repair and overhaul facility requirements in the following detail:

(1) Space requirements for all repair/overhaul functions.
(2) Loading.
(3) Power and other utilities.
(4) Environmental conditions.
(5) Special furnishings exclusive of S&TE and documentation.

(f) *Table C-6, Intermediate Supply Support Facility.* Table C-6 shall specify the requirement for supply support facilities developed in the Facilities Plan in the following detail:

(1) Space requirement.
(2) Loading requirements.
(3) Environmental conditions.
(4) Power and other utilities.
(5) Security requirements.
(6) Special furnishings.

4. *Appendix D – Depot Requirements.* This appendix shall contain five tables listing the assemblies of the system which are repairable at the depot level, and the documentation, support and test equipment, facilities, and personnel required to accomplish these repairs. Each of the five tables is discussed below:

(a) *Table D-1, Depot Repairable Item List.* Table D-1 contains the list of assemblies which are repairable at the depot and will include the following additional description for each assembly.

(1) *Reference designation.* Circuit designation of the assembly in the equipment.
(2) *Nomenclature.* Name of the assembly.
(3) *Part No.* Part number of the assembly.
(4) *Total Qty/system.* Quantity of this assembly type in the equipment.
(5) *Maintenance requirement No(s).* Maintenance requirement No(s) from applicable worksheet which necessitates depot repair.

(b) *Table D-2, Depot Documentation.* Table D-2 shall list all of the documentation required by the depot to accomplish disassembly, repair/overhaul, assembly, test, inspection and calibration of the modules and test equipment. The list information shall be keyed to the equipment by reference description.

(c) *Table D-3, Depot Support and Test Equipment.* Table D-3 shall list all of the support and test equipment required by the depot to accomplish repair/overhaul. The following data shall be provided for each item listed:

(1) Name.
(2) Part No.

(3) Quantity required.
(4) Price.
(5) FSN (as required).

(d) *Table D-4, Depot Personnel.* Table D-4 contains the skill levels required by the depot personnel to accomplish disassembly, test, repair/overhaul, assembly and calibration of each module. In addition, the list shall include an approximation of the number of assemblies per year which require depot repair and any additional training which will be required by the depot personnel.

(1) *Training requirements.* Any unusual training requirements needed by depot personnel to accomplish the repair.

(e) *Table D-5, Depot Repair Overhaul Facilities.* Table D-5 will specify the depot level repair and overhaul facility requirements in the following detail:

(1) Space requirements for all overhaul/repair functions.
(2) Loading requirements.
(3) Power and other utilities.
(4) Environmental conditions.
(5) Special furnishings exclusive of S&TE, documentation, etc.
(6) Storage.

5. *Appendix E – Maintenance Task Requirements Summary*

(a) Maintenance tasks are the expanded statements of maintenance re-uirements and consist of step-by-step procedures to be followed by the technician when performing preventive or corrective maintenance at each level of maintenance.

(b) Maintenance tasks will be so described and sequenced that the maximum economy will be realised in carrying out the process they collectively describe for a specific maintenance requirement or set of requirements. The anticipated frequency of occurrence will also be specified for each task.

BIBLIOGRAPHY

American Society for Quality Control (1975). *Proceedings, Annual Reliability & Maintainability Symposium Research/Development Reliability.*

Anonymous (1986). *Production Engineering*, **33**, (3), pp. 68–70, March.

Bazarsky, Igor (1961). *Reliability Theory and Practice* (Englewood Cliffs, NJ: Prentice-Hall, Inc.)

Beskin, Clay (1986). 'Is JIT Spelled Barcoding?' *Manufacturing Systems*, **4**, (10), October.

Blanchard, Benjamin S. (1986). *Logistics Engineering and Management*, 3rd edn. (Englewood Cliffs, NJ: Prentice-Hall, Inc.)

Bowersox, Donald J., Closs, David J. and Helferich, Omar K. (1986). *Logistical Management*, 3rd edn. (New York: Macmillan).

Brooks, K. and Watsman, A. (1986) 'Garbage Disposal: to burn or not to burn', *Chemical Week*, 11 June, **138** (34), pp. 44–47.

Constantinides, A. (1986). *Basic Reliability.*

Coyle, John J., Bardi, Edward J. and Langley, C. John Jr (1988). *The Management of Business Logistics* (St Paul, MN: West Publishing Co).

Defense Systems Management College (1986). *Integrated Logistics Support Guide*, May.

Defense Systems Management College (1986). *Systems Engineering Management Guide*, 2nd edn, Dec.

Downs, William R. (Downs Technical/Management Services), *Report on Basic Maintainability.*

Garske, Scott and Murray, Kathleen (1988). *Hospital Material Management Quarterly*, **9**, (4), pp. 35–40, May.

Goldman, A.S. and Slattery, T.B. (1964). *General Reference Maintainability* (New York: John Wiley & Sons, Inc).

Heizer, Jay and Render, Barry (1988). *Production and Operations Management: Strategies and Tactics* (Boston: Allyn and Bacon, Inc).

Krajewski, Lee J. and Ritzman, Larry P. (1987). *Operations Management: Strategy and Analysis* (Reading, MA: Addison Wesley).

Leenders, Michiel R., Fearon, Harold E. and England, Wilbur B. (1985). *Purchasing and Materials Management*, 8th edn. (Illinois: Richard D. Irwin, Inc).

Page, Harry R. (1980). *Public Purchasing and Materials Management* (Lexington, MA: Lexington Books).

Raytheon Company. *Life Cycle Costing*. Lexington MA.

Sherman, Stanley (1985). *Government Procurement Management*, 2nd edn. (Gaithersburg, MD: Wordcrafters Publications).

Taff, Charles A. (1984). *Management of Physical Distribution and Transportation*, 7th edn. (Homewood, IL. Richard D. Irwin, Inc).

Temple, Barker and Sloane, Inc (1982). *Transportation Strategies for the Eighties* (National Council of Physical Distribution Management).

Twiner, William (1985). *Secrets of Personal Persuasion* (New York: Prentice Hall).

US Department of Defense. *Integrated Logistics Support Guide*, DOD Directive DoD-D-4100.35, 1964.

US Department of Defense. *Major System Acquisitions*, DOD Directive 5000.1.

US Department of Defense. *Major System Acquisition Procedures*, DOD Instruction 5000.2.

US Department of Defense. *Acquisition and Management of Integrated Logistic Support for Systems and Equipment*, DOD Directive 5000.39.

US Department of Defense. *Test and Evaluation*, DOD Directive 5000.3.

US Department of Defense. *Reliability Program for Systems and Equipment Development and Production*, MIL-STD 785.

US Department of Defense. *Maintainability Verification/Demonstration/Evaluation*. MIL-STD 470.

US Department of Defense. *DOD Requirements for a Logistic Support Analysis Record*. MIL-STD-1388-2.

US Department of Defense. *Maintainability Prediction* MIL-STD-472. 1984.

US Department of Defense. *Technical Reviews and Audits for Systems, Equipments, and Computer Software*, MIL-STD 1521. 1978.

Zenz, Gary J. (1981). *Purchasing and the Management of Materials* (New York: Wiley & Sons).